Nature's Curiosity Shop

Nature's Curiosity Shop

Barry E. Zimmerman
and
David J. Zimmerman

CONTEMPORARY
BOOKS
A TRIBUNE NEW MEDIA COMPANY

Library of Congress Cataloging-in-Publication Data

Zimmerman, Barry E.
    Nature's curiosity shop / Barry E. Zimmerman and David J.
Zimmerman.
        p.   cm.
    Includes index.
    ISBN 0-8092-3656-7 (paper)
    1. Science—Miscellanea.   I. Zimmerman, David J.   II. Title.
Q173.Z556   1995
500—dc20   94-47076
                                                                    CIP

Published by Contemporary Books, Inc.
Two Prudential Plaza, Chicago, Illinois 60601-6790
Manufactured in the United States of America
International Standard Book Number: 0-8092-3656-7
10   9   8   7   6   5   4   3   2   1

To our wives, Marilyn and Sondra—
two writers' widows who never wavered
in their support of our dream

To Amy, Tara, and Corie, our beautiful daughters—
who never wavered in their anticipation
of our royalty checks

Contents

Acknowledgments

Thanks to William and Margaret Rabus, Margaret Colvin, Carol Courtney, Bonnie Savitz, and Fred Miller for their support and research assistance.

Introduction

I do not know what I may appear to the world; but to myself I seem to have been only like a boy playing on the seashore, and diverting myself in now and then finding a smoother pebble or a prettier shell than ordinary, whilst the great ocean of truth lay all undiscovered before me.

Isaac Newton spoke these words more than 250 years ago. Indeed, he found many smooth pebbles and pretty shells, as did the scientists who came before and after him. This book takes a closer look at some of those pebbles and shells and explores "the great ocean of truth" that is nature.

Nature's Curiosity Shop is a collection of twenty-eight essays on some of the most fascinating subjects in biology, chemistry, astronomy, physics, and earth science. Among the topics explored in these pages are eclipses, rain forests, sex, time travel, and Einstein's theories of relativity. You will learn, among other things, why scientists have trouble predicting the weather and why but for a bend in a molecule of water there would be no life on Earth.

We live in an increasingly technological society, one of

supercomputers and complex information networks, of genetically engineered organisms and virtual reality. Advances in science are measured in days and weeks rather than years. How can one keep pace?

This book is a good place to begin. Its purpose is to make science accessible to everyone. It was written to inform and entertain, to explain without becoming tedious or trivial. But don't take our word for it—move on to the next page and find out for yourself.

Virtual Reality: A World of Make-Believe

When you look into a mirror, the image you see of yourself appears to be inside. There is, of course, nothing in the mirror. Physicists call this mirror representation a *virtual image*, *virtual* meaning "not real" or "imaginary." Scientists today have reached far beyond the mirror to create the imaginary. Using powerful computers, they are building entire worlds that appear to be real but are not. When you step into one of these worlds, you enter a *virtual reality*.

The term *virtual reality* was coined by Jaron Lanier, founder of VRL Research, in 1989, which tells you how new the field is. As defined by Steve Aukstakalnis and David Blatner in their book *Silicon Mirage*, it is "a computer generated, interactive, three-dimensional environment in which a person is immersed." But how can a computer generate this artificial world? More intriguing, how can it immerse you in this world and allow you to interact with it? The same way elephants give birth—with difficulty.

The Technology

The objective of a computer programmed for virtual reality is to fool you into believing you are within a world that does

1

not exist. To do this it must trick your senses, specifically your sense of sight, hearing, and feeling. Of the three, visual deception is the most critical and the one that has received the most attention from those developing the technology.

Today virtual realities are created by head-mounted displays—rather cumbersome helmetlike affairs that cover the eyes. Built into the helmet is a 2- to 3-inch screen for each eye. In the better-designed systems the screens wrap partially around the head so that your peripheral vision confirms that the virtual world exists.

The screens can be either cathode ray tubes (CRTs) or liquid crystal displays (LCDs), both types familiar to us. The CRT is a typical television screen or computer monitor. LCDs are lighter in weight and cheaper, but CRTs produce higher-resolution pictures. In fact, LCD graphics are often cartoonlike—not very realistic.

Whatever the screen type, once the computer is doing its thing, images called stereo pairs flash rapidly and continuously before the viewer's eyes at a rate of sixty or more per second. Each eye receives a different image at a slightly different time; the effect is to generate a three-dimensional environment in which the viewer is instantly immersed. But the fun has just begun. Also constructed into most helmets are sensors that perform a necessary function known as *position/orientation tracking*. They follow every movement, every turn, every nod of your head and convey this information to the computer.

There are several kinds of tracking devices, and each has its advantages and disadvantages. One type has a camera on the top of the helmet that picks up different flashing ceiling lights as you move, informing the computer as to your whereabouts. In Nintendo's popular PowerGlove game system, players wear a specially constructed glove that emits high-frequency clicking sounds. Microphones in the Nintendo computer detect these ultrasonic clicks and translate them into the on-screen movement of a character's fist.

A third tracking method, and the most popular, is called *magnetic position sensing*. Two sets of wire coils, at various angles to one another, are built into a helmet or glove. Electric current passes through one set of coils, producing a small magnetic field. As the hand or head is moved, the second set of coils moves through the magnetic field, inducing a tiny electric current in these coils. A properly programmed computer can use this induced electricity to track the movement of the particular body part. VPL's Data-Glove uses this type of motion detection. It is vastly superior to Nintendo's PowerGlove, and it costs about a hundred times as much.

The widespread use of magnetic position sensing notwithstanding, it is *image extraction* that will be the tracking method of the future. Using a series of video cameras that follow your movements from a number of different angles, image-extraction technology then feeds the data into a computer programmed to interpret this movement. It is the most accurate of all position-sensing methods, producing the most realistic on-screen movement. Better still, image extraction does not require the wearing of cumbersome equipment. Unfortunately, it also is the most calculation-intensive of all methods. Its dependence on powerful supercomputers and very complex programming will probably keep it off the consumer market until well into next century.

Accurately tracking your movements in the virtual world is essential, for only then can the computer properly alter the environment. When you turn to the right, you expect to see the scenery move as it would in the real world. When you walk across the room, approaching objects should become larger. When you walk around a chair, not only should the appearance of the chair change smoothly and appropriately, but it should exhibit parallax, moving across your field of view faster than distant objects. And if a lamp is lighting the room, the computer must constantly adjust shading and shadowing to create the illusion of reality. Needless to say, all this altering and adjusting is computationally very demanding. To avoid the jerky, choppy movements of a motion picture playing in slow motion, the computer must refresh or update a scene at least fifteen times a second. And the more complex the environment—the more objects the computer must deal with—the more calculating it must do. Today's computers are just not powerful enough to work that fast with that much data. So virtual worlds are still constructed simply and inhabited by figures that are caricatures of real people.

As if the computer wasn't having enough trouble fooling your eyes, it also has to contend with your ears. To complete the deception, the three-dimensional virtual world must be filled with sound that is indistinguishable from what you would hear in a similar real environment. A radio blasting in a corner of the room must sound as if it is coming from a corner of the room. As you move farther away, the sound must grow fainter. If you move to a different part of the room, you must hear it coming from a different angle.

In their book, Aukstakalnis and Blatner describe an interesting experiment performed by Fred Wrightman and Doris Kistler of the University of Wisconsin. Wrightman and Kistler wanted to determine exactly what the ear hears as it picks up sounds from three-dimensional space. This data could then be fed into a computer, which could mimic the shape of sound waves as they impinge on the ear from different directions. They started by seating a subject in a room and placing a tiny probe microphone deep within the subject's ear, near the eardrum. Different tones were played, one at a time, through each of 144 speakers situated throughout the room. Each tone from a particular speaker was picked up by the microphone and recorded. Remarkably, when the recording was played back through headphones, the tones sounded as they did when originally played through the different speakers.

With enough data from experiments such as this one *and* a set of stereo headphones *and* a position/orientation tracking hookup *and* a supercomputer loaded with extraordinarily complex algorithms (problem-solving strategies), sound that approaches real life and changes in appropriate real time can be achieved.

That brings us to the sense of touch. When you grasp a virtual ball, turn on a virtual light switch, run your fingers through your virtual girlfriend's hair, dip your hand into a virtual bucket of water, or touch the virtual thorn on a virtual rose, you want some feeling, some sensation that remotely resembles the real thing. To exist in a virtual environment without even rudimentary tactile sensations would greatly diminish the illusion. Unfortunately, until very recently people have had to settle for this greatly diminished illusion. Tactile feedback simply was not available. Only now are strides being made, and the results still fall far short of reality. Reproducing the sense of touch represents, perhaps, the greatest challenge to VR researchers.

There is a fundamental problem. Unlike seeing and hearing, tactile experiences require that contact be made between your skin and something material. Most efforts to provide this sort of stimulation have concentrated on the hands, which are, after all, our principal manipulators of the environment, containing our greatest concentration of tactile receptors. It works something like this: Through the use of specially designed gloves, every movement of the hand, including bending of the fingers, is tracked. This is then mimicked by a computer-generated virtual hand on a three-dimensional head-mounted display. So far, so good. But what happens when the virtual hand reaches out and grasps an object? The glove must convey the sense of having touched something.

It does this in one of several ways. As far back as the mid-1980s, for example, the U.S. Air Force was sewing piezoelectric crystals into the fingertips of gloves. These crystals vibrate when subjected to an electric current, giving the sensation of touch. Quite crude and inexact, it was nevertheless used with success by pilots on virtual instrument panels during flight simulations.

A second and perhaps more ambitious attempt to bring sensation into the virtual world involves the use of many tiny air bladders, scattered throughout the underside of a glove. When the proper bladders are inflated, pressure is applied to the fingers and palms when you handle a virtual object. Once again, not very true to the touch, but what can you expect when you reach out and grasp something that isn't really there?

Not very much—at least for the foreseeable future. Although the situation will improve as human ingenuity comes up with ever more creative ways of stimulating nerve endings and muscles of the hand as well as the entire body (VPL has already created a DataSuit to go along with its DataGlove), there are definite limits to what can be accomplished. For instance, you will never be able to sit in a virtual chair or lean on a virtual wall. There is nothing there to support you. Even experiencing the weight of an object as you lift it will be impossible unless the glove is secured to a device that can offer resistance. But half a loaf of tactile stimulation is better than none. Let's see just what sort of worlds have been created today, and will be created tomorrow, with our half a loaf.

The Application

Virtual reality won't merely replace TV. It will eat it alive.

Arthur C. Clarke

In this statement one of our greatest science fiction writers expresses, rather dramatically, the belief that VR will affect our lives most profoundly in the areas of entertainment and recreation. But is this just a bit of hyperbole? Probably not. Although VR will never match the complexity or richness of texture found in the real world, it comes close enough to offer the magic of the two Is—*immersion* and *interaction*. By contrast, television is purely passive. There is no way a VR user could ever be described as a couch potato. The action is too fast and furious. Take Virtual Racquetball, for example, an arcade-style game created by Autodesk. By donning a helmet, slipping on a DataGlove, and grabbing a racquet with a motion-tracking system, you suddenly find yourself on a virtual racquetball court, ready to return serve. All movements of your racquet are reproduced faithfully by computer-generated graphics as you play against either the computer or another person, who can be thousands of miles away. As the game progresses, the adrenaline begins to flow. By game's end, you've worked up quite a sweat.

Dactyl Nightmare, by W Industries, is yet another VR interactive game that has been operating in about a dozen virtual reality centers in the United States as of March 1993. The object of the game is to seek out and shoot your opponent—another player—before he finds and shoots you. An additional threat is posed by a computer-generated pterodactyl, which swoops down and tries to grab you with its claws. It is a crude simulation, as are all VR game systems at this point. The virtual world, although stereoscopic, is cartoonlike. There is no tactile feedback, and you move about by simply pressing a button on a joystick handle. Nonetheless, people are queuing up to pay for the experience—at a cost of $4 for just 3½ minutes. Perhaps it should be called "Financial Nightmare."

There is little doubt that VR will figure very prominently in the future as an entertainment medium. It can make fantasies come true. Ease into your helmet, grab a specially constructed guitar, and the computer will have you performing at a rock concert, in front of thousands of screaming fans. You'll never go back to karaoke again.

Recreational uses of VR will be limited only by humankind's imagination. And its popularity, if home entertainment systems such as that of Nintendo are any indication, will grow exponentially. It should be pointed out, however, lest we view VR as nothing more than a "virtual boob tube," that there is more to virtual reality than mindless arcade games. Its true value is not in creating artificial worlds but in helping us better understand the real one. For example, at the University of North Carolina in Chapel Hill chemists are slipping on goggles and immersing themselves in a three-dimensional world of twisted and contorted protein molecules. Using joysticks or control arms, they maneuver representations of smaller drug molecules in an attempt to dock them into the protein's active sites. There is even resistance and attraction built into the joystick—the electromagnetic forces of the protein. As the drug molecule approaches, the protein is distorted in unpredictable ways. Through these simulations, pharmaceutical companies hope to learn before actual manufacture begins which experimental medications will work best.

Computer simulation of reality: According to a 1992 article in *Science*, it is being called the "third branch of science" (the article never explains what the first two are). NASA is using it at its Ames Research Center to re-create the surface of Mars based on data gathered from Martian satellites and probes. Urban developers are employing an eighty-block by eighty-block virtual model of riot-torn southern Los Angeles to help rebuild the area. By walking or flying through the simulation, they can note and subsequently correct oversights and errors in design. In much the same way, and for much the same reason, architects and automobile manufacturers are using VR to make mock-ups of new home and car designs. *Science News*, in January 1992, reported, "In Japan a customer in an appliance showroom can design a new kitchen on the computer screen, then don goggles and glove to wander through the imaginary custom kitchen's array of gleaming gadgets."

Virtual reality of sorts (although not truly immersive or interactive) has even been employed in a courtroom. One particular case involved the 1991 murder of a San Francisco porno-movie king, allegedly by his brother. A virtual crime scene, with reenactment of the murder, was created by the prosecution. After viewing the simulation, the jury came back with a guilty verdict.

It is interesting to note that the military is investing very heavily in VR research and is, without question, the largest source of funding for virtual reality research and development. "Were it not for the Defense Department," say Aukstakalnis and Blatner, "virtual reality would probably not exist." Using state-of-the-art headgear and bodygear and the world's fastest supercomputers, the military is creating almost-real environments in which to train pilots to fly. Through SIMNET, the world's largest immersive simulation, military personnel throughout the United States and Europe are engaging in war games and simulated terrorist attacks, preparing for the day when they will not be simulations. Between 1992 and 1996 the United States military plans to spend more than $500 million on simulations. (One helmet for a flight simulation can cost up to a million bucks.)

Medicine is also getting into the act. Doctors are now practicing surgical procedures on virtual patients with various and sundry virtual injuries, tumors, etc. Consider the following scenario as described by Aukstakalnis and Blatner in *Silicon Mirage*:

> Reaching out, Dr. Miyakawa could feel the virtual scalpel in his glove hand, and when he cut the skin, tiny computer-controlled motors silently whirred, creating a resistance that could only be described as lifelike. Retraction of the skin, blood flow, internal anatomy, and potential problems were all carefully programmed into the computer and displayed (through goggles) to the surgeon as he progressed.

Jaron Lanier, a pioneer and true giant in the field of VR, feels that medicine will be virtual reality's "monster market" in the future. Someday a doctor on one part of the globe will be able to perform delicate operations on a patient thousands of miles away. Cameras on a robot in the operating room will be the eyes of the surgeon. Gloves worn by the doctor will not be of the sterile rubber kind but rather tracking and tactile-feedback affairs that remotely control, with infinite precision, the surgical implements gripped by robotic hands.

Experiencing real yet remote environments, such as an operating room many miles away, is called *telepresence*. Purists argue that telepresence is not truly virtual reality since an artificial, computer-generated environment has not been created. They are correct, but why split hairs? The technology, the hardware, the immersion and interaction are all there. And certainly the interest is there. The next century will find businesspeople scattered around the world, hooking into computers, slipping on head-mounted displays, and coming together in virtual conference rooms to meet and discuss whatever it is that businesspeople are wont to discuss. Although oceans apart, they will be immersed in a simulation wherein they interact as if truly assembled in the same room. It is called *teleconferencing*, and it is an idea whose time has come.

Are you afraid of heights? Snakes? Are you arachnophobic? Well, you're in luck. Therapists are planning to use VR to cure you of your neuroses. By having patients confront virtual snakes and spiders, they hope to desensitize them and eventually effect a cure.

Sex. VR has some good times in store for you in that area as well. Unfortunately, by the time realistic virtual sex has finally been developed you might be too old to enjoy it. According to Howard Rheingold, writer of the bestselling book *Virtual Reality*, virtual sexual encounters are "an early-to-mid-twenty-first century technology, rather than next year's fad." The reason is twofold: lack of computing power and lack of proper hardware (no pun intended) to create the necessary tactile sensations. But if the following Rheingold description of futuristic virtual sex is any indication of things to come, it might be well worth the wait:

> Before you climb into a suitably padded chamber and put on your 3-D glasses, you slip into a lightweight (eventually one would hope, diaphanous) bodysuit, something like a body stocking, but with the intimate snugness of a condom. Embedded in the inner surface of the suit, using a technology

that does not yet exist, is an array of intelligent sensor-effectors—a mesh of tiny tactile detectors coupled to vibrators of varying degrees of hardness, hundreds of them per square inch, that can receive and transmit a realistic sense of tactile presence. . . . You can run your cheek over (virtual) satin and feel the difference when you encounter (virtual) flesh. Or you can gently squeeze something soft and pliable and feel it stiffen under your touch.

Now imagine plugging your whole sound-sight-touch telepresence system into the telephone network. You see a lifelike but totally artificial visual representation of your own body and of your partner's. . . . You run your hand over your partner's clavicle, and 6,000 miles away, an array of effectors are triggered in just the right sequence, at just the right frequency to convey the touch exactly the way you wish it to be conveyed.

Not bad! The simulated sexual experience is called, aptly enough, *teledildonics*, and it beats the heck out of teleconferencing. Perhaps, after your virtual tryst, you can flip a computer toggle and take a refreshing virtual cold shower. Then reach for that thick and thirsty virtual bath towel and dry yourself off. Unfortunately, when the electronic magic is over and you slip out of your bodysuit, you'll still be hot and coated with nonvirtual sweat. C'est la vie!

My Brother, the Clone

I am a *clone*. My creation was not, however, the result of some ungodly scientific research. I am a clone by virtue of the fact that I have an identical twin. He is, in fact, the coauthor of this book.

Very simply, clones are genetic duplicates. When I was conceived, the egg from which I developed did not remain one mass of cells as it divided after fertilization. Early on it split into two clumps. The clump destined to grow into an intelligent, handsome, charismatic young fellow became me. The other became my twin brother. But the important thing is that we came from the same *zygote*, or fertilized egg. (Identical twins are more correctly called *monozygotic twins*.) This means our DNA (our genetic makeup) is identical. Twinning is a kind of natural cloning.*

Cloning is commonly encountered among the species. Whenever organisms reproduce asexually—without a partner—clones of the single parent result. Virtually all plants clone themselves. The tubers (potatoes) of potato plants, the

---

*The nine-banded armadillo is interesting in this regard. Females always give birth to four identical twins because the fertilized egg splits into four parts during cell division.

runners of strawberry and spider plants, the bulbs of tu-
lips—these all give rise to clones of the parent. Seeds, in
contrast, are the result of pollen grains fertilizing eggs—
sexual reproduction. Each seed germinates into a genetically
unique individual—a nonclone.

Cloning maintains the sameness of a particular variety
of plant or animal from one generation to the next. It is a
feature of asexual reproduction that humans have been taking advantage of for a long time. When you make a cutting
of your favorite houseplant and root it in soil, the resulting
plant is a clone of the original. When you eat a Delicious or
a McIntosh apple or a seedless orange, you're eating a clone,
one created by grafting a twig or slip of the desired fruit
tree onto an ordinary host tree. The term *clone* is, in fact,
derived from the Greek word *klon*, meaning "twig" or
"slip." Specific varieties of grapes, bananas, sweet potatoes,
sugarcane, pineapple, asparagus, and even garlic have been
artificially cloned as far back as 4,000 years ago.

Animals are not as universally asexual as plants.
Whereas all animals reproduce *sexually*, asexual *cloning* is
common only among the simpler invertebrates. A jellyfish or
flatworm will pinch off a tiny bit of itself, and that mass of
cells will grow into a new individual. The female greenfly
produces eggs that develop, without fertilization, into
greenflies genetically identical to their mother. A few species
of fish, amphibia, and lizards can also perform this cloning
trick, known as *parthenogenesis* (see "Sex Among the Species" for more on parthenogenesis). But by and large vertebrates are incapable of growing a new individual from a part
of themselves, be it a large chunk or a single cell. Animals
lost that ability once they evolved into vertebrates. Can
science come to their aid? Should science come to their aid?

Animal Cloning: A History

The science of animal cloning is really a study of cell differentiation. Why can a fertilized egg cell divide into cells that
become bone and muscle and skin, whereas an adult skin cell

can only make more skin cells, and an adult muscle cell only more muscle cells? Why can't a skin cell, which has all the DNA of a fertilized egg cell, do what an egg cell can do— make a new human being? Therein lies the greatest unsolved mystery of biology. When we figure out the riddle of embryonic development, we will have the keys to curing cancer and a host of other human afflictions as well as halting the aging process. We will know how genes work.

One of the earliest theories intended to explain cell differentiation, proposed by German biologist August Weismann just before the turn of the century, suggested that genetic determinants of the zygote were divided and subdivided among the embryonic cells at each division. Eventually, each cell would get the genetic determinants necessary for that cell's specialization. A skin cell, for example, would get determinants only for skinness and a muscle cell determinants only for muscleness.

This theory did not hold up under scientific scrutiny and was subsequently discarded. When a two-celled frog embryo was shaken so that the two cells separated, each cell developed into a healthy tadpole and then a healthy adult frog. Determinants were not being divvied up among the cells.

We now know that every cell in every division gets all the genetic material of the original zygote. What happens is this: somewhere along the way, genes in cells get irreversibly shut down, limiting the potential of the cells to differentiate. As embryonic development proceeds, different genes turn off in different cells, locking them into their own particular fates. Can cells be unspecialized? Can the genes that have shut off be made to switch back on again? Can a skin cell ever regain, through intervention of science and technology, its potential to create a whole human being? The answer is no, at least not yet, although it makes for very interesting science fiction.

Ira Levin, in the 1976 book (and subsequent movie) *The Boys from Brazil*, had neo-Nazi scientists taking Hitler's body cells and growing them into new Hitlers. As if Hitler weren't bad enough, writer Michael Crichton, in *Jurassic Park*, cloned dinosaurs from scraps of dinosaur DNA. The necessary genetic material was recovered from the bellies of prehistoric insects that had fed on dinosaur blood before being fossilized in amber. These books were admittedly science fiction, novels written solely to entertain. In 1978, however, a book by David Rorvik called *In His Image—The Cloning of a Man* hit the bookstores. It asserted that human cloning was already a fait accompli—that a wealthy businessman had paid scientists over $1 million to transform one of his skin or liver cells (the book is unclear as to which cell was used) into a duplicate of himself.

Ultimately the book was denounced as a hoax by scientists around the globe. One even successfully sued the publisher for unauthorized association of his name with the alleged cloning. Rorvik did, however, paint a very convincing portrait of the human cloning process, drawing on research done on lower vertebrates such as frogs.

Ah, the poor frog. Along with the sea urchin, a small marine creature related to the starfish, it has been pursued mercilessly by embryologists since the latter part of the nineteenth century. One reason is the size of frog eggs—huge by mammalian standards. Another is their external fertilization and development, ideal for manipulation and study of differentiating embryos.

So, with frog eggs in hand—or, rather, in a dish of pond water—embryologists sought to clone an adult frog. The procedure was simple:

1. Remove the nucleus with its genetic material from the body cell of an adult frog.
2. Place it in an egg cell, activated by contact with a sperm, that had had its nucleus removed through microsurgery.
3. Watch the zygote develop into a clone of the frog that donated its nucleus.

This type of research, known as a *nuclear transfer*

experiment, was destined to fail because the adult cell nucleus, even though it was in a fertilization-activated egg, could not mobilize its shut-down genes and differentiate into a new organism. So, in the early 1950s the experiment was tried again, this time using the nucleus from a cell of a day-old frog embryo. At one day of age the embryo is called a *blastula* and consists of several thousand cells. The experiment worked, and development proceeded normally, resulting in a healthy tadpole and adult frog.

Amazingly, when the nucleus was taken from a *cancer* cell of an *adult* frog and placed in an enucleated egg, a tadpole and then adult frog resulted as well. This came as a shock to many biologists, who were expecting nothing more than a mass of tumor cells to develop. Obviously, when cells turn cancerous, genes are switched on that should remain shut down in a specialized adult cell. Perhaps in the future the same cancer cells that are killing you can be used to clone a *new* you. An interesting thought, but let's get back to reality.

In the mid-1970s Dr. J. B. Gurdon and his associates at Oxford University were able to take the nucleus from an intestinal cell of a *fully developed tadpole*, transfer it into an enucleated egg cell and, once again, produce a normal frog. Obviously the genes for differentiation had not yet been irreversibly suppressed at the tadpole stage, and the hospitable environment of an egg cell was able to switch them back on. Unfortunately, the mechanisms involved are still a mystery.

Work with frogs has resulted in several major advances. But what about mammals? Has any sort of cloning success been realized in furry, warm-blooded creatures like us?

Yes. In the early 1980s biologists in Geneva cloned three mice. They did so in much the same way that frogs are cloned. Nuclei were taken from cells of very early embryos and transplanted into enucleated eggs activated by fertilizing sperm. Unlike fertilized frog eggs, however, the mouse embryos had to be placed back into a female mouse for development.

Since those salad days of embryology there have been no new earth-shattering breakthroughs on the cloning front. Adult cell nuclei still resist the coaxing and cajoling of prying biologists and simply refuse to click on the genes that would unspecialize them. In late 1993 there was a to-do in the media over an apparent cloning of a human being, but it was overreaction. No adult human being had been cloned. Scientists hadn't even come close. Twinning maybe. Blastomere separation for sure. But no cloning. Let's take a closer look at what was actually accomplished.

Blastomere Separation

When a fertilized egg begins to divide, the resulting cells of the embryo are called *blastomeres*. In 1970 two blastomeres were separated by animal researchers and used to produce identical twin mouse pups. Any womb can gestate the developing embryos, and it need not be that of the biological mother. In 1979 blastomere separation at the Institute of Animal Physiology in Cambridge, England, created several sets of identical twin lambs. Soon the procedure was being employed to create calves. Among animal breeders seeking to produce many offspring from superior livestock, this sort of "artificial twinning" is old hat. But in October 1993 Jerry L. Hall and Robert J. Stillman performed the first successful blastomere separation in humans. By "successful" I do not mean that they actually implanted the developing blastomeres in human wombs. Such experiments would be ethically and legally out of the question. What they did was this:

Working in George Washington University's in vitro fertilization clinic, Hall and Stillman selected seventeen embryos that had been fertilized by more than one sperm. Such defective embryos do not survive and were slated for the scrap heap. All of those chosen were in stages of early cleavage (divisions following fertilization) and consisted of from two to eight blastomeres. An enzyme, pronase, was then employed to dissolve the gelatinlike coating called the *zona pellucida* that normally protects the developing egg. Blastomeres were separated and wrapped individually in a synthetic zona derived from seaweed.

The team allowed these blastomeres to develop in a dish. Cells from the two-celled embryos did best, some even making it to the thirty-two-celled stage. At this point they could have been implanted into wombs. All in all, Hall and Stillman produced forty-eight "clones."

The science was nothing new. As already mentioned, it had been going on in the livestock business for a dozen years. Hall and Stillman had not even transferred nuclei from embryo cell to egg cell, standard procedure among cattle breeders. So why the fuss? Because humans are not cattle. And this was the first experiment performed on a human, albeit nonviable, embryo.

Human embryo research does not sit well with many people, who envision sinister uses for the technology. Baby farms could be created, using the sperms and eggs of "superior" human beings to create "superior" blastomeres. Signs such as "wombs for rent" would dot the landscape. A family might be created with identical twins of different ages. Merely freeze separated blastomeres until such time as another twin is desired. Taken to its illogical conclusion, a woman could conceivably give birth to her identical twin. If a baby turned out to be superior in any way, one might even be able to sell its frozen clones to the highest bidder.

Ridiculous? Most biologists believe so, giving such futuristic scenarios about as much credence as cloning an army of Hitlers. Yet consider this: in 1990 Abe and Mary Ayala admitted they had conceived a baby girl to provide bone marrow for an older sibling dying of leukemia. Suddenly the notion of freezing embryos to be grown and cannibalized for spare body parts at some later date does not seem so fantastic. And clones, because they are perfect tissue matches of their genetic doubles, would encounter none of the problems of organ rejection.

Small wonder, then, that medical ethicists condemn any form of human experimentation, be it cloning or twinning, with an almost religious fervor. There is, however, an upside to early embryo experimentation. For starters, such research will benefit thousands of infertile couples. Presently, in vitro fertilization enjoys only a 10 percent success rate when doctors work with a single egg. These figures would rise dramatically were blastomere separation employed to produce several implantable embryos.

Blastomere separation would also find widespread application in a new form of genetic diagnosis. Using the procedure, a couple with a family history of a serious genetic disease (Tay-Sachs, cystic fibrosis, or muscular dystrophy, for example) would undergo in vitro fertilization. A micropipette would then punch a hole in the zona pellucida of the embryo and suck out a blastomere, whose DNA would be checked for the defective gene. The trouble is, sometimes one cell is not enough. If the single blastomere could be multiplied by Hall and Stillman's techniques, enough DNA could be gathered for a successful test.

The bottom line is this: we cannot learn if we do not permit ourselves to question and to investigate. Early surgeons could not learn their craft without cutting open human bodies. Scientists cannot learn how fertilized eggs differentiate without studying fertilized eggs. It is as simple as that. Perhaps we should heed the words of the great seventeenth-century English philosopher Thomas Hobbes, who wrote "knowledge is power." We can only hope such power will be used to benefit humankind.

How Cold Is Cold?
How Hot Is Hot?

Place a thermometer in your mouth for several minutes; then remove it. The temperature should read 98.6°F. If you're not feeling well, it might read 103°F or 104°F. Most of us know that if it reads 110°F, we're probably dead. If it reads 90°F we've died some time ago. We understand these values in terms of body temperature. We also understand temperature as it is used in weather reports. At 20°F, for example, you put on a winter coat when you venture outdoors; at 90°F, you might go to the beach. We are familiar with temperature as it relates to our everyday lives. But what exactly *is* temperature? What is different about air at 20°F and 90°F—or 9,000°F?

Temperature: A Matter of Degree

Temperature is a measure of how hot or cold something is. It is determined by the motion, or *kinetic energy*, of molecules.

Perpetual Motion

All molecules move: they may fly off in all directions, as in a gas; or stick together but slide over one another, as in a liquid; or remain fixed in place but vibrate, as in a solid. It is this movement that creates heat and, in turn, temperature. The greater the kinetic energy of a group of molecules, the hotter is the temperature. Since molecules, even within a small sampling of matter, move at different speeds, temperature is defined as *the average kinetic energy of a group of molecules*. If you are in good health and do not have a fever, the temperature of your body, as already noted, is 98.6°F— or very close to it. But what does the 98.6 stand for? In ounces, 98.6 is a bit more than six pounds. In inches, it is a bit more than eight feet. What is 98.6°F?

Just as inches and feet are units on a scale for measuring distance and ounces and pounds are units on a scale for measuring weight, °F are units on one of several scales used to measure temperature. It is called the *Fahrenheit scale*. Daniel Gabriel Fahrenheit was a German-Dutch physicist who, in 1714, built the first truly accurate thermometer. Unlike earlier thermometers, which were based on the expansion and contraction of gases such as air or liquids such as alcohol or water, Fahrenheit's thermometer used liquid mercury. It proved superior to other substances for a variety of reasons. First, it is the most uniform among liquids in its rate of expansion; water, in particular, expands in a *very* nonuniform way with temperature change.* Second, mercury doesn't freeze as easily as water does or boil as easily as alcohol does—it remains a liquid through a rather wide temperature range. Third, water and alcohol both cling to the surface of glass. Thus, as the temperature of a water or alcohol thermometer dropped, some of the liquid remained behind, clinging to the glass and slowly trickling down. This made it exceedingly difficult to get accurate temperature readings.

Fahrenheit also devised a temperature scale with which

---

*This anomalous behavior of water is discussed in "The Miracle Molecule."

to calibrate his acclaimed thermometer. It was based loosely on a suggestion by Isaac Newton that the fixed points of a thermometer be the freezing point of water and the temperature of the human body, with the freezing point of water marked at 0°. Fahrenheit realized that this would cause the temperature on many winter days to be *below zero*. In an effort to avoid negative temperatures as much as possible, he added salt to water, which lowered its freezing point. The temperature at which the salt water froze he designated 0° Fahrenheit (or 0°F). He then divided the interval between this point and human body temperature into 96 equal parts (don't ask me why), making human body temperature 96°F. With this scale the freezing point of pure water was *close to* 32°F and the boiling point *close to* 212°F.

As scientists began to learn more about water's physical and chemical properties, the freezing and boiling points of *pure water* assumed greater importance as reference values. In response, Fahrenheit adjusted his scale so that its freezing point was *exactly* 32°F and its boiling point was *exactly* 212°F, with the interval between divided neatly into 180 equal parts. This reset the temperature of the human body at 98.6°F.

The Fahrenheit scale is used primarily in the United States, by nonscientists. Most other countries (even Great Britain, where the Fahrenheit scale originated) and scientists *everywhere* use another scale, devised by Swedish astronomer Anders Celsius in 1742. His scale was marvelously simple. It placed the freezing point of pure water at 0°, the boiling point at 100°, and divided the interval between into *one hundred* equal parts. (Curiously, Celsius first placed the freezing point at 100° and the boiling point at 0° but reversed it the following year.) At first it was called the *Centigrade scale*, meaning "hundred steps," but since 1948, by general agreement, it has been called the *Celsius scale*. (In either case the symbol *C* was used as the degree designation, so there was no need to revise the textbooks.)

A comparison of the Fahrenheit and Celsius scales is given in Figure 1.

Figure 1

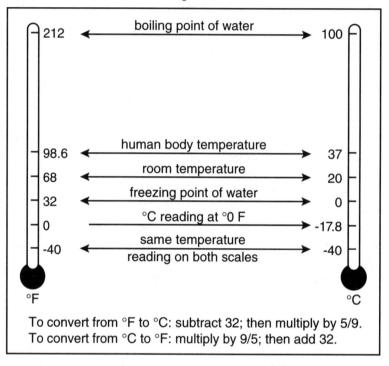

To convert from °F to °C: subtract 32; then multiply by 5/9.
To convert from °C to °F: multiply by 9/5; then add 32.

Another temperature scale was the Réaumur scale, devised by the French physicist René-Antoine Réaumur in 1731. It marked the freezing point of water at 0° and the boiling point at 80°. This scale held its own against those of Fahrenheit and Celsius for a while but slowly lost favor and is now little more than a scientific curiosity.

After Réaumur not much happened in the world of temperature scales for more than a century. There seemed little reason to divide the interval between the freezing and boiling points of water into yet another set of equal parts. But then came developments in the world of physics that made the freezing point of water an arbitrary and quite meaningless starting point for temperature scales.

How Low—or High—Can You Go?

We already know, and have known for thousands of years, that as things get hot they expand and as they get cold they contract (with a few notable exceptions, such as water at certain temperatures; see "The Miracle Molecule"). Solids and liquids expand and contract very little with temperature change, and the rate at which they do so varies widely from substance to substance. For example, the metal aluminum expands seventy-seven times faster than the mineral quartz. This is not true for gases. All gases expand and contract very much with temperature change, and they do so at the same rate—about three hundred times faster than the average solid.

The constant rate at which gases contract raised an interesting question. If a gas contracts at a steady rate as it cools, getting smaller and smaller, wouldn't it eventually contract to nothing? What then? How can matter have no volume, take up no space . . . in essence, disappear?

French chemist Joseph-Louis Gay-Lussac addressed this problem in the latter part of the eighteenth century.* He took a given volume of gas at a starting temperature of $0°C$ and cooled it slowly, recording volume change. For every degree he cooled the gas (*any* gas), the volume lost $\frac{1}{273}$ of its original volume. For example, if he started with 273 gallons of oxygen gas at $0°C$, when he lowered the temperature to $-1°C$, the volume contracted to 272 gallons. At $-10°C$ the volume became 263 gallons; at $-50°C$ it became 223 gallons; and so on. What if he lowered the gas temperature to $-273°C$? Would its volume contract to zero?

Not really. When gases are cooled enough, they become liquids, and their rate of contraction slows down greatly. At even lower temperatures the liquids solidify, and their rate of contraction slows down even more, so the volume might

---

*He was not so worried about expansion. It is easy to accept that a confined gas might expand indefinitely as it gets hotter; there is certainly room enough in the universe.

never drop to zero. But the work of Gay-Lussac and others led to another interesting question: was there a limit to how cold things could get?

The number −273°C remained an intriguing value. As physicists began to link temperature change to kinetic energy of molecules rather than volume change, they reasoned that not only did the oxygen gas in our experiment lose $\frac{1}{273}$ of its volume for each degree of drop in temperature from 0°C, but the molecules lost $\frac{1}{273}$ of their *motion*. At −273°C the molecules would have no motion. Even the electrons that move within the atoms that make up the molecules would have no motion. Total kinetic energy would be zero. This did not sit well with physicists. If electrons did not move around inside atoms, they would be pulled into the center, or nucleus, by their mutual attraction, and the atoms would collapse. All matter would collapse at −273°C.

To make a long story short, matter cannot be allowed to collapse; therefore, temperature cannot be allowed to reach −273°C. It is indeed a lower temperature limit. Even with today's advanced technology we have never reached −273°C (−273.16°C, to be exact) and, in all likelihood, never will. It is one of those "unreachables" in the physical world, which, if it ever were reached, would cause very strange and exotic things to happen. The speed of light is another unreachable. (See "It's All Relative: The Special Theory.")

Reachable or not, −273°C was quickly put to good use. In 1848 Scottish physicist William Thomson (later raised to the rank of baron and the title Lord Kelvin) suggested that the starting or zero point on a temperature scale should not be the freezing point of water (as on the Celsius scale) or 32° below (as on the Fahrenheit scale), but the lowest temperature that can exist. (After all, water is just one of countless substances, such as alcohol, apple juice, and chocolate mousse.) It made good sense. He called this lowest temperature *absolute zero* and the scale that derived from it the *absolute scale*. It is more often called the *Kelvin scale*, however, in his honor.

For his purposes Kelvin chose to modify the Celsius

rather than the Fahrenheit scale. One degree Kelvin (1 K) and 1°C are actually equivalent in value.* (If the temperature today is 10°C warmer than it was yesterday, it is also 10 K warmer.) The only difference is the starting point. Kelvin starts at 0, Celsius at −273. (In case you're wondering, Fahrenheit starts at −460.)

The Kelvin scale is preferred in certain fields of science, such as astronomy and physics. Table 1 compares some common values on both the Kelvin and Celsius scales.

Table 1
Going Absolute

	°C	K
boiling point of water	100	373
human body temperature	37	310
room temperature	20	293
freezing point of water	0	273
absolute zero	−273	0

To convert from °C to K, add 273.
To convert from K to °C, subtract 273.

We've learned how low temperature can go. Is there an upper limit as well?

−460°F to 11,000°F
(−273°C to About 6,000°C)

Although we've never reached absolute zero, we have come awfully close—a few billionths of a degree above. At this temperature there are no gases and almost no liquids. The air you breathe is a solid block of ice. At the molecular level everything is moving in superslow motion.

---

*With the Kelvin scale the degree symbol (°) is usually omitted.

As things heat up, molecules begin to shake and shimmy. Many break the bonds that are holding them together. Melting and boiling become the order of the day. Pressure plays a part in the physical state of matter at different temperatures. In most cases for the melting and boiling points of substances the pressure is standard, or roughly atmospheric pressure at sea level.

At two Fahrenheit degrees or one Celsius degree above absolute zero (we'll stick with these two commonly used scales from now on), helium melts; a few degrees above that (−452°F or −269°C) it boils. It is the first element to do so. Hydrogen follows close behind. Oxygen is fourth on the list, melting at −361°F (−218°C) and boiling at −297°F (−183°C). If we warm up another 168°F, we reach the coldest weather temperature ever recorded on Earth: −129°F (−89°C). It happened in Antarctica, in 1983. The first metal to melt is mercury, at −38°F (−39°C). Water, as we all know, melts at 32°F (0°C).

We are now on the positive side of the temperature scales and moving steadily upward. In Libya in 1992 the hottest shade temperature on Earth was recorded, a sweltering 136°F (58°C). But that's cool compared to boiling water: 212°F (100°C). And *that's* cool compared to a kitchen oven, which typically produces an air temperature of up to 550°F (288°C). This is somewhat higher than the temperature at which book paper burns. In fact the title of a novel by Ray Bradbury, *Fahrenheit 451*, is derived from paper's kindling point. The story takes place in an oppressive futuristic society, where the government burns books to suppress freedom of thought and expression.

The air inside an oven doesn't get nearly as hot as the flame that heats it. A natural gas (methane) flame in a standard oven or on a stovetop burner reaches temperatures in excess of 2,000°F (1,093°C). That's about the melting point of gold and copper and the temperature a car's exhaust can reach. (The car's engine gets much hotter—about 4,500°F.) At this temperature heated objects will cast a reddish glow.

Many other metals melt at higher temperatures—iron, for example, at 2,795°F (1,535°C). The metal with the highest melting point is tungsten, the filament that glows inside your light bulbs. It becomes a liquid at 6,116°F (3,380°C). The only element that melts at a higher temperature is diamond, which begins to splash at a temperature of 6,400 °F (3,538°C). Curiously, tungsten *boils* at the highest temperature of all the elements, even higher than diamond: 10,700°F (5,927°C).

Above this temperature—in fact, above 7,300°F (about 4,000°C)—no solids exist at standard pressure.

But things get a lot hotter than this. Where do these superhigh temperatures occur, and what processes create them?

11,000°F (About 6,000°C) and Above

If you go to the center of the Earth—some four thousand miles beneath your feet—the temperature is about 12,000°F (6,649°C), a bit higher than on the surface of the Sun. The material is mostly metallic iron and nickel, in solid form because of the extremely high pressure. It is high pressure that also creates the high temperature. Larger and more massive planets, which create greater internal pressure, have even hotter cores. Jupiter, the largest and most massive planet in the solar system (more massive than the other eight planets put together), has a core temperature estimated as high as 36,000°F (about 20,000°C). This temperature is so high that all molecular bonds fail. Existing substances are composed of individual atoms or parts of atoms.

Now that we are reaching temperatures beyond those generated by the planets, let us consider the stars. The surface temperatures of stars vary greatly. The smaller ones are actually rather cool—5,400°F (about 3,000°C) or so. They look red. Our Sun, which is about ten times more massive than small stars and a thousand times more massive than Jupiter, builds up significantly greater internal pressure and heat. The temperature at its surface is roughly 10,000°F (about 5,500°C), and its glow is yellow. But the Sun is far from the most massive star—and therefore far from the hottest. At the surfaces of the most massive and hottest stars, temperatures can reach 200,000°F (about 111,000°C) and the stars glow blue-white. Laser beams, incredibly, can get about *ten times hotter* than this.

But lasers and star surfaces are no match for temperatures in the *centers* of stars. Even the smallest stars have core temperatures in excess of 18,000,000°F (about 10,000,000°C). For that is the point at which *hydrogen fusion* occurs—the process that powers all stars and causes them to shine. At its center the Sun reaches a temperature of about 27,000,000°F (about 15,000,000°C). The centers of heavier (more massive) stars get even hotter. On Earth, scientists duplicate hydrogen fusion reactions when they explode H-bombs. Experimental nuclear reactors called *tokamaks* also duplicate hydrogen fusion under controlled laboratory conditions. (Do not confuse tokamaks with the nuclear reactors that are used worldwide to produce energy for society's needs. These are *fission* reactors, which use an entirely different nuclear process, release somewhat less energy, and are not at all experimental.) The temperature inside tokamaks may get as high as 720,000,000°F (about 400,000,000°C).

Can things get hotter than this? Yes. When stars use up their hydrogen and begin to die, they go a little crazy. They begin to fuse other elements, and their core temperatures may rise well above 2,500,000,000°F (about 1,400,000,000°C). The heaviest stars may end their life in an incredible explosion, called a *supernova*. Temperatures in the core of a star going supernova may exceed 41,000,000,000°F (about 23,000,000,000°C). In particle accelerators temperatures in excess of 1,800,000,000,000°F (about 1,000,000,000,000°C) have been reported.

So we know that temperatures can reach trillions of degrees. Where does it all end? Physicists feel that as there is a lower limit to temperature, so there is an upper limit.

According to the article "High Temperature" by John Hastie and David Bonnell of the National Institute of Standards and Technology, and Joan Berkowitz, of Risk Science International, in Washington, DC, "The maximum realizable temperature in nature, limited by the onset of [matter] creation from the available energy, has been estimated to be of the order of 2×10^{12} K." That's 3,600,000,000,000°F (about 2,000,000,000,000°C)—about twice as hot as anything we know of today.

Except for one thing: the Big Bang! That is how we believe our universe started, and the temperature at which this event must have occurred is mind-boggling. Cosmologists and astrophysicists have calculated the Big Bang temperature to have been on the order of 200,000,000,000,000,000,000,000,000,000,000°F (change the 2 to a 1 and you're close enough in °C). Nothing gets hotter than this. Nothing.

The Miracle Molecule

It's why your heart beats.
It's how your blood flows.
It helps your lungs breathe, your joints move,
your muscles work.
It's your body's cooling system.
   Excerpt from an Evian bottled-water commercial

And it's much, much more. It's water.

Water is the most widely used of all substances. Without it there would be no life, climates around the world would be vastly different, and, most important, food would not be microwavable. What makes water so special among substances? The key lies in the structure of its molecule.

The Proper Angle

For all its complex behavior and uniqueness, water is a rather small and simple molecule. It consists of two hydrogen atoms bonded separately to an oxygen atom, as shown in Figure 1.

Figure 1

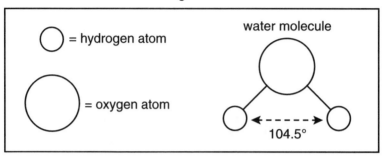

The three atoms in a water molecule form an angle of 104.5 degrees. The bond between each hydrogen atom and the oxygen atom is formed by the sharing of a single pair of electrons and is called a *covalent bond*. It is very important to note, however, that the electrons are not shared equally. Oxygen is much hungrier for electrons than hydrogen is (a property known as *electronegativity*). In other words, when a hydrogen atom bonds with an oxygen atom, the electrons that are shared in the process spend much more time around the very electronegative oxygen atom. As a consequence the oxygen side of the bond becomes negatively (−) charged and the hydrogen side becomes positively (+) charged. (This is because electrons are negatively charged, in case you've forgotten your high school science.)

These charges would be insignificant, however, if water were a linear molecule. Such a molecule would be symmetrical (see Figure 2).

Figure 2
The Way It *Isn't*

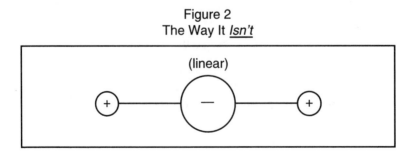

(To test for symmetry of a three-atom molecule such as water, draw a horizontal and vertical line through the middle of the molecule. Top to bottom, as well as left to right, it will look the same, if the molecule is symmetrical.) The positive charges would be distributed evenly around the negative charge and have a canceling effect. There would be only one *center of charge*; the molecule would not have a negative and positive side, or *pole*. It would be a *nonpolar molecule*.

But water is *not* linear; it is angular. And that makes all the difference (see Figure 3).

Figure 3
The Way It *Is*

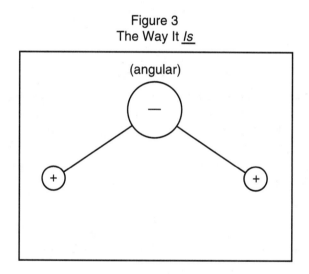

Being angular, the molecule is not symmetrical. The positive charges are not distributed evenly around the negative charge. They do not cancel out but instead have their own center of charge. The molecule has a positive and a negative pole. It is a *polar molecule*, what chemists call a *dipole* (see Figure 4). In fact, water is a *very* polar molecule—more so than almost any other molecule.

So water molecules are not linear but angular. But for the slight bend, life would not exist.

Figure 4
Water: A Polar Molecule, or Dipole

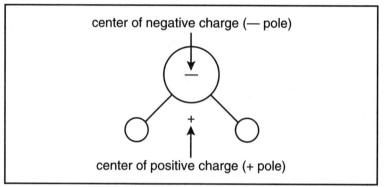

center of negative charge (— pole)

center of positive charge (+ pole)

Sticking Together

Because dipoles have negative and positive poles, they behave like little magnets. The positive pole of one molecule attracts the negative pole of an adjacent molecule. This causes the molecules to stick together. Honey is sticky for this reason. Figure 5 illustrates this attraction between water molecules.

Figure 5
Dipole-Dipole Attraction

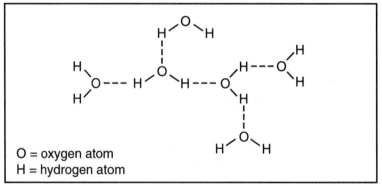

O = oxygen atom
H = hydrogen atom

The attraction is represented by a broken line, and since it involves hydrogen atoms, it is known as a *hydrogen bond*. (A hydrogen bond is an attraction *between* polar molecules that contain hydrogen. Hydrogen bonds are weaker than the covalent bonds between oxygen and hydrogen atoms *within* the water molecule.) Water molecules, because of their distinctly dipole nature, are held together by strong hydrogen bonds. They have a powerful tendency to stick to each other. How powerful? Using a medicine dropper, carefully place drops of water onto a penny. Count how many drops the penny holds before the water spills over the edge.

To demonstrate the stickiness of water molecules another way, fill a glass or cup with oil and another with water. Holding it as flat (parallel to the surface) as possible, carefully place a small paper clip on the surface of each liquid. Voilà! The paper clip floats on water but not on oil. Yet the clip is made of steel—a substance much denser than either water or oil. It should sink in both.

Actually the clip is not floating on water for the usual reason—buoyancy, or a difference in density. Because of water's stickiness, molecules at the surface hold together, forming a kind of skin that resists penetration. It is called *surface tension*. If you push the paper clip below the surface, it will indeed sink.

Molecules that stick together form solids and liquids. A great deal of energy must be added to water, usually in the form of heat, to unstick its molecules and cause them to fly apart—to transform it into gas. In other words, water has a high boiling point. It likes to be a liquid. The temperature of water must be raised to 212°F (100°C) before its molecules have enough energy to overcome their strong hydrogen bonding and fly apart. Between 32°F (0°C) and 212°F, water is a liquid. In almost all regions of the world water exists in the liquid phase for at least part of the year.

If, however, water molecules were linear rather than angular, and therefore nonpolar rather than polar, they would not be sticky, and water's boiling point would be drastically lower. Estimates are that water would boil at −85°F (−65°C) if it were nonpolar. It would be a gas at just about all temperatures that exist on Earth.

It is the fact that water is a liquid—angular and polar rather than linear and nonpolar—that makes it the wellspring (a watery metaphor) of life.

Elixir of Life

Life on Earth came about through a complex molecule-building process called *chemical evolution* (see "How Did Life on Earth Start?"). This process involved the mixing and reacting of many different compounds. There is no better substance than liquid water for this purpose: for dissolving materials and providing a medium in which they can bump into each other and react. Water has been called the *universal solvent*. It is particularly good at dissolving the many substances that are a part of the living world. Without liquid water, life certainly could not have evolved. That it originated in the seas and thrived there for hundreds of millions of years before invading the land is ample proof.

Earth, where life has thrived, is aptly called a *water planet*. Liquid water covers nearly three-quarters of the planet's surface, weighing in at about 1½ quintillion (1,500,000,000,000,000,000) tons. (That's 28 septillion—or 28,000,000,000,000,000,000,000,000—drops!)

Other plants are not so lucky. Mars is a dry world and therefore lifeless. The presence of erosion channels on its surface indicates, though, that water may once have flowed there. That means life may once have existed. The Moon is lifeless because, among other reasons, it is bone-dry. Ditto for Mercury. Venus is lifeless because it is too hot for liquid water. Lead would melt on its surface. The gas giants are just that—gas giants. Pluto is an orbiting ice cube. Forget about life on Pluto.

Not only was liquid water essential to the origin of life; it is also essential to the maintenance of life. (If you need proof, try drinking a glass of water vapor.)

A living organism has been described, somewhat tongue in cheek, as "a bag of enzymes." What this means is that a living thing is a complex, highly structured machine that must accomplish thousands of different chemical reactions at a time. Enzymes coordinate and speed up these reactions. Without enzymes, many would occur too slowly. Water is the medium of choice for the working of these enzymes. No other liquid could successfully dissolve as many substances and allow them to react. No other liquid comes even close, in fact. Perhaps that is why water is the most abundant chemical in living things. It makes up 70 percent to 85 percent of all cells. You and I are 60 percent water by weight; our brains are 70 percent. Even our bones contain water—20 percent by weight. The average 150-pound (68-kilogram) person lugs around about forty quarts (thirty-eight liters) of water.

Water also flows easily, unlike the viscous liquids oil and molasses. This feature, along with its superior dissolving capabilities, makes water an excellent transport or circulatory fluid. Blood, which is about 93 percent water, dissolves nutrients and hormones, as well as metabolic wastes, carrying them to and from the body's cells.

Water also carries off fairly large amounts of heat when it evaporates. When you are overheated and sweating, the water in the sweat evaporates from your skin, carrying off heat, cooling you. Water is your body's coolant.

In fact it is an excellent temperature stabilizer, because it has a high *specific heat*—it does not heat up or cool down very easily. Water helps to keep your body temperature constant—at around 98.6°F (37°C). The temperature-stabilizing effect of water has important applications outside the living world as well.

Water and Climate

Land heats up and cools down much more quickly than water. It has a lower specific heat. This causes land areas to have a greater temperature range from season to season than coastal areas. For example, Reykjavík, the capital of Iceland, and Verhoyansk, Siberia, have about the same elevation and latitude. They receive about the same amount and intensity of sunlight from day to day. You would, therefore, expect their climates, especially temperature, to be similar. Not so. Reykjavík is on the southern coast of Iceland, by the Atlantic Ocean. It has an annual temperature range of only 20°F (11°C). Verhoyansk, on the other hand, is deep in the interior of the continent of Asia. Its annual temperature range is *120°F (67°C)!* The temperature-moderating effect of water can be evidenced in the United States as well. Inland cities such as Omaha, Nebraska, have far greater annual temperature ranges than coastal cities such as Los Angeles, California.

An Anomaly of Water

Most substances become less dense as they are heated. (It is easiest to view density as the compactness of matter or the extent to which its molecules are squeezed together.) Take a copper penny, for example. As the penny is heated, the atoms of copper in the penny move faster and spread apart. The penny takes up a bit more space; its density decreases. If you continue to heat the penny, it will eventually melt. As a liquid the density of copper is even less than as a solid. Heating the liquid copper causes its molecules to continue spreading apart, and it becomes less and less dense. This pattern is true for almost all pure substances—but not water.

Take water at 50°F (10°C): it is a liquid. Instead of heating it as we did the copper, we will cool it. As the liquid is cooled, the water molecules move more slowly and come closer together; the density of the water increases, as expected. But at 39°F (4°C) an odd thing happens—the water molecules begin to spread apart as the substance is cooled further. At 32°F (0°C), the water freezes, and its molecules *spread apart even more*—by nearly 10 percent! (This is why antifreeze must be added to a car's radiator in regions where temperatures drop below "freezing." If the water were allowed to freeze, it would expand and crack the car's engine block.) In other words, water at 39°F is denser than water at 32°F. It is, in fact, as dense as liquid water ever gets. And liquid water at *any* temperature is denser than ice. This is why ice cubes float in a glass of liquid water. This odd behavior is due to the fact that, as ice, water molecules form a rather *open* crystal arrangement. In melting, this open structure collapses, allowing the molecules to come closer together, increasing the density of the substance. The open structure is not completely collapsed until water reaches 39°F.

Water's unusual behavior has some interesting effects in the world around us. Take, for example, the metamorphosis undergone by a lake or deep pond as the seasons change. As winter approaches and the air temperature drops, water at the surface of the lake cools, becomes denser, and sinks, causing warmer water from below to rise and be cooled in turn. Beyond 39°F, however, cooling does not cause the water to sink. As it cools from 39°F to 32°F, it becomes less dense and remains at the surface. Eventually it will freeze and turn to ice. *A body of water freezes from the top down.* Nearly all other liquids would freeze from the bottom up.

Freezing from the top down allows water in the lake or pond to remain liquid despite an air temperature below 32°F. The ice at the surface serves as a thermal barrier, insulating the liquid water below from the frigid air above. Usually, except in a shallow pond, the bottom of the body of water remains liquid. This enables marine inhabitants to survive severe winters.

A Flaky Situation

Water can also be a work of art. In the solid phase it may take a fascinating and quite beautiful form: *snow*. The thought of snow often brings to mind unpleasant thoughts—hours of shoveling, immobile automobiles, frostbite—yet a snowflake is one of nature's most delicate designs. The next time it snows, catch a few flakes on a sheet of black paper and examine them closely with a magnifying lens. You will notice that they are all six-sided, or *hexagonal*—a consequence of the manner in which the water molecules connect. If the air temperature where they formed was relatively warm (though below freezing), the flakes will be larger and more complex. If the temperature was cold, they will be smaller and simpler. This is because warmer air is generally moister and provides more water molecules for the growth of snow crystals.

These remarkable crystals vary greatly in their patterns. They may be flat with internal designs, like a hexagonal piece of embroidery; or columnar, with six-sided cylinders; or star-shaped, with six-sided branches radiating from the center in different directions. The shapes of snow crystals are determined largely by temperature, but other factors— even the rate at which they fall to Earth—influence their structure.

The structure of a snowflake is very open; that is, snow crystals contain many large holes. As a result, snow has a low density—much lower even than ordinary ice (which, as you already know, is lower than liquid water). In fact, *more than fifty inches* of dry, powdery snow would melt down to about *one inch* of rain.

Snow forms in the upper atmosphere as the result of water vapor collecting and solidifying around motes of dust. It does not usually result from liquid water freezing, which is interesting.

It has been said that no two snowflakes are exactly identical. Is this actually true? In one respect it is. The average snow crystal contains approximately ten quintillion (10,000,000,000,000,000,000, or 10^{19}) water molecules. The different ways they can arrange themselves in three dimensions is nearly infinite. No two flakes will have exactly the same molecular configuration. But in a broader sense many snowflakes have the same look and size. To quote meteorologist Jack Williams of *USA Today* in his *The*

Weather Book, "Many tiny snow crystals are simple, hexagonal plates with no obvious difference in shape. Even more complicated crystals can be alike."

Good Vibrations

Foods cook in a microwave oven because they contain water. Remember, water is a strong dipole; electrically, it is a molecule with a very positive and a very negative end. Microwaves are a type of electromagnetic radiation, like radio waves, light, and x-rays. They have electrical and magnetic properties associated with them. When they pass through food, they cause polar molecules, such as water, to vibrate, or oscillate, rapidly. These oscillations generate heat. It is this heat that cooks food in a microwave oven. Microwave ovens do not exactly cook food from the inside out, as is sometimes claimed, but the source of cooking heat is within the food rather than outside it. Hence microwaved foods do not brown or crisp.

The story of water is indeed the story of a miracle molecule.

Your mind needs it to think, and your eyes to see.
It's most of what you're made of.
And drinking enough of it every day is one of the easiest, most important ways to keep your body healthy for a lifetime.

Excerpt from Evian commercial

Sex Among the Species

All living things engage in sex. Male frogs can be seen in early spring, at pond's edge, riding females, whom they cling to tenaciously with the "nuptial pads" of their thickened thumbs. As the female releases her eggs into the water, the male ejects his milky sperm. Although the male has no penis to introduce into the female, he does not seem to mind the union. Some species of small tree frogs will remain clamped in copulation for as long as six months! (I guess female frogs don't get headaches.)

Bees unwittingly aid and abet floral promiscuity by delivering pollen from one flower to another on their hairy bodies. The flower is the organ of sexual reproduction in higher plants, and pollen is to the plant world what sperm is to animals.

Even microscopic one-celled pond dwellers such as slipper-shaped paramecia can, with a bit of patience, be seen pressed up against one another in what would appear to be a passionate embrace. Although I am not certain of the passion, it *is* a sexual encounter, for there is transfer of genetic material.

Indeed, *all* living things engage in sex. But why such ubiquity? There is, after all, a simpler and easier way to procreate. It is called *asexual reproduction*, and it was around a billion years before the sexual variation on the theme ever evolved.

Sexual vs. Asexual Reproduction

In asexual reproduction only one parent is involved. Many, if not most, plants can reproduce this way. When you make a cutting of your favorite houseplant, for example, you are reproducing it asexually. When bacterial and other one-celled organisms split in half, producing two daughter cells, they are also reproducing asexually—without benefit of a partner. Fungi, such as molds and mushrooms, often produce spores—asexual, reproductive cells that fly off and grow into new molds and mushrooms.

Even animals can be found reproducing asexually, although the practice is not very widespread and usually is restricted to the lower taxonomic groups. Sponges, for example, the simplest of all animals, merely shed clumps of cells that readily grow into new sponges. Hydra, a quarter-inch-long aquatic animal related to the jellyfish, grows a new hydra right out of the side of its body. This unusual and asexual means of propagation is called *budding*.

Sea anemones, another relative of the jellyfish, are sessile animals that grow on the ocean floor. Their brightly colored tentacles give them the appearance of underwater flowers. When they want to reproduce, sea anemones simply split themselves down the middle, thus creating two new anemones.

Certain flatworms have the remarkable ability to pinch off a piece of themselves whenever they feel like starting a family. The small, darkly colored planarian, often seen clinging to the underside of submerged logs and rocks, is such a worm.

As we move up the evolutionary ladder, asexual reproduction becomes less and less common. Starfish do not usually reproduce asexually, having a pair of testes or ovaries in each of their arms. They can do so, however, with the aid of unsuspecting clam farmers. Starfish eat clams. They slowly pull open the shell with their rays or arms and then drop their stomachs out of their mouths and onto the clam's succulent viscera. Clam growers do not appreciate the starfish's penchant for clams on the half shell. When they catch a starfish in their nets they cut it into several pieces and throw it back into the ocean. This is not a wise move. They assume that they have destroyed the starfish. Yet each piece, if it retains a portion of the central disk, can regenerate lost parts and grow into a new starfish—it can reproduce asexually.

Among the vertebrates—fish, amphibia, reptiles, birds, and mammals—asexual reproduction is rare and not very spectacular. An arm or a leg does not break off and proceed to grow into a new organism. When asexual propagation does occur, an egg merely develops into a new individual without benefit of fertilization by a sperm. The phenomenon, called *parthenogenesis*, is found among certain species of fish, amphibia, and lizards. In parthenogens such as the Amazon molly, a brilliant little fish of Texas and Mexico, males simply do not exist. Some people believe that parthenogenesis, or virgin birth, took place at least once in the human species—a birth that founded Christianity.

That's about it as far as asexual reproduction goes among animals. Despite the ease and simplicity of propagation without a partner, insects, spiders, crabs, lobsters, and earthworms reproduce almost exclusively through male-female copulation, as do the vertebrates. Despite the valuable time and energy expended in hunting down a mate, despite the need for complex reproductive organs and structures, sexual reproduction is universal. Even the simplest animals, those that reproduce asexually as well, often opt for sex instead.

Why? What is the great advantage of sexual reproduction that it has caught on so? In a word, variety. When organisms reproduce asexually, they make new individuals that are carbon copies of themselves. They not only look alike: they are genetically identical (except for an occasional mutation). Exact likeness in an entire population of organisms is not good. If all the fish in a lake were identically sensitive to a certain fungus, then an entire population of fish might be wiped out when the fungus struck. If however, there is variation in the population—if individuals are more or less sensitive to the fungus—then some fish will survive to perpetuate the line.

This is the gift of sexual reproduction: *diversity* within a population. When a male and female copulate, there is a mixing of genetic material. The new individual receives a generous helping of genes from each parent, no two individuals receiving quite the same set of genes—not even two individuals from the same parents. The only exception is monozygotic twins, which are genetically identical (see "My Brother, the Clone" for more on monozygotic twins).

Finding a Mate and Delivering the Goods

Sexual reproduction involves finding a partner. This poses a major problem for creatures living in harsh and sparsely populated environments, such as the cold and dark of the deep sea. How is a deep-sea male anglerfish to mate if he rarely encounters a female? More important, what is he to do for a prom date? Have no fear. The male angler has developed a novel and quite remarkable way of dealing with this problem.

Male anglerfish are very small, about half an inch long, compared to the four feet of some females. This extreme *dimorphism* originally led taxonomists to group them into separate species. At an early stage in his development the male angler clamps himself to a female. So secure is the attachment that the male literally grows into the female, even fusing his circulatory system with hers. The miniature male stops feeding and becomes totally parasitic. He is, in effect, nothing more than a copulatory organ, reading the hormonal changes in the female's bloodstream and releasing clouds of sperm on demand. Not terribly macho behavior, but it works for the angler.

Barnacles are small, feathery creatures that hide within crusty shells. Unlike anglerfish, they do not lead lonely existences in the inhospitable depths of the ocean bottom. Quite the contrary. Great numbers of barnacles are found growing in shallow marine waters. They do have a problem, however—a sessile way of life. Barnacles permanently attach to rocks, wood pilings, ship hulls, and the bodies of other, larger marine animals. They are firmly cemented by their heads, and this inability to move to seek out a mate puts a definite crimp in the barnacle's love life.

But the barnacle also has found a way around its dilemma, as can be witnessed by its reproductive success. On the Isle of Man, estimates of one billion barnacles have been reported in a thousand-yard stretch of beach. Thirty thousand barnacles often crowd into a square yard of seashore. This success is the result of two adaptations—*hermaphroditism* and a huge penis.

Proportionally speaking, the barnacle possesses the largest copulatory organ of any animal—twice as long as the barnacle itself. It unfurls this 1½-inch monster and introduces it within an opening of the shell of a neighboring barnacle. Since barnacles grow in densely packed groups, this doesn't usually pose a problem. And the barnacle does not have to concern itself with the gender of its neighbor. Every barnacle is a hermaphrodite, possessing both male and female sex organs (testes and ovaries) and producing both sperm and eggs. This means that every barnacle is capable of inseminating and of being inseminated by every other barnacle. And if the neighbors are all more than a penis length away, it is believed that barnacles can produce young through self-fertilization. Where there is a will there is a way.

Hermaphroditism is not at all uncommon in the animal world. Many worms are hermaphroditic, as are mollusks such as the snail and slug. Perhaps a more interesting setup is that of the blue-headed wrasse, a fish that inhabits coral reefs. Male wrasses have blue heads and are larger than the females. A dominant male will preside over a harem of about a half dozen females, mating with each one daily. When the male wrasse dies, the largest female, over a two-week period, grows a blue head and male sex organs and becomes a functioning male. In another type of wrasse, the cleaner wrasse (so named because it eats external parasites from other fish), a female can turn into a male in a matter of hours. Even the large butterfly shrimp you order in fancy eating establishments are jumbo females that earlier in their lives were small, insignificant males.

The need to find a mate will drive some animals to fairly desperate measures. A case in point is the hay itch mite. Mites are tiny (sometimes microscopic) eight-legged critters belonging to the spider class, Arachnida. Many are parasitic, feeding on blood, skin, or other animal tissues. The hay itch mite is such a parasite. It leads an extremely insular existence, never leaving its host in search of mates. To a mite the particular animal it parasitizes is an isolated, solitary island. In the absence of available paramours and with a compelling need to procreate, the hay itch mite does the unthinkable—it commits incest. Adrian Forsyth, in his book *A Natural History of Sex*, aptly describes the action:

> [Hay] mites are born as adults. All the interme-
> diate stages have been compressed into one. The
> first individuals to hatch are males. They take on
> the role of obstetrician. They insert their pincerlike
> legs into their mother and begin pulling out their
> sisters, and as soon as a male gets possession of his
> sister, he mates with her. . . . A male may proceed to
> inseminate all of his twenty or so sisters.

In another type of mite the male actually begins inseminating his sisters while all are still inside the mother's
womb. He then dies without ever having been born. His

nonvirginal sisters promptly cut a hole through their mother's abdomen and crawl away.

There are also oedipal variations on the incest theme, with virgin females laying *unfertilized* eggs that develop into male mites. Maternal restraint not being what it should be, the mother soon succumbs to the charms of her newly hatched sons and mates with them. The clutch of *fertilized* eggs that she subsequently lays all develop into females, which wander off to repeat the bizarre incestuous pattern in which, as Forsyth says, "the son is husband to his mother and father to his sisters. The sister is the daughter of her brother and the wife of her son." Sound like a soap opera?

One of the more unusual methods of sperm transfer is demonstrated by our somewhat perverse friend the bedbug. Bedbugs are insects, the only invertebrate group of animals most of whom possess a penis. They are very tiny insects that hide during the day in cracks in the furniture and floors, emerging at night to sink their mouthparts into unsuspecting sleepers. After gorging themselves with blood, they depart, leaving itchy, red blotches as their calling cards. Not very neighborly. Then again, neither is raping other male bedbugs, something bedbugs do rather shamelessly. There is, however, a reason for this homosexual assault. The sperm ejaculated into the rape victim will travel to his sperm duct. When he next copulates with a female, he will inseminate her not with his own sperm but with the rapist's.

Even when bedbugs engage in heterosexual sex, they do so with panache. Although the female of the species has a perfectly normal reproductive system, the male bedbug chooses to stab the female in her abdomen with his sword-like member. The ejaculated sperm migrates to the female's circulatory system and then to a storage organ, eventually fertilizing a batch of eggs that she will lay. In the meantime, a special pad of abdominal tissue mends her puncture wound. You can't make this stuff up.

Every bit as bizarre as the bedbug's mating habits, are those of the segmented marine worm *Platynereis megalops*.

(Segmented worms also include earthworms and leeches.) When gathered in swarms, the females proceed to bite off the tail segments of the males. Contained within these segments are mature sperm. Digestive juices in the female liberate the sperm, which then penetrate the walls of her gut, enter the body cavity, and somehow find and fertilize her eggs. (Luckily, the male Platynereis is capable of regenerating his lost segments.)

Male honeybees are not as fortunate. During mating the male honeybee literally explodes, his genitalia detonating like a hand grenade. Why, you might ask, would any creature evolve such a self-destructive means of copulation? Evisceration toward what end? Strange as it might appear, the explosive nature of his coital suicide drives his reproductive apparatus deep within the queen, where it lodges firmly. This, in effect, plugs her up, ensuring that his sperm and none other will do the fertilizing.

As already mentioned, many insects have a penis. It helps them overcome the sperm-transfer difficulties encountered by a terrestrial way of life. Spiders are also terrestrial animals, but they do not have penises. Not to worry. The spider has developed a pair of ventral appendages called *pedipalps* into copulatory organs that work much like eye-droppers. Suction bulbs near the tips of the palps collect sperms and then discharge them directly into the female's genital opening. Unfortunately, during or immediately after mating, our eight-legged Casanova is often eaten by the female. Expendability of the male spider, especially if he provides nutrition for his mate, should surprise no one. (It's how the female black widow got her name.) Once he has performed his appointed task, the male is unnecessary. As Forsyth so eloquently stated, "An individual is merely a gene's way of transmitting itself."

Another fascinating sexual adaptation among animals are *spermatophores*. They are solid packets of sperm, and they are produced by quite a number of insects, worms, mollusks, and even salamanders and newts. The leech—a blood-sucking, pond-dwelling segmented worm—has a spermatophore that he cements onto the back of a female. The spermatophore coat contains an irritant, which creates an ulcer on the female's skin. Sperm then enter the female through this ulcer and travel to the ovaries, where ripe eggs await. The spermatophore of the giant octopus is pencil-thick and more than a yard long. Snails produce weirdly shaped spermatophores that match the internal twists and turns of the female's vagina. Some spermatophores are huge affairs, weighing up to half the male's bodyweight. Whatever its size or shape, the spermatophore is a multipurpose structure. It can plug up the female's reproductive opening, much as the honeybee's bursting genitalia does, effectively eliminating competition from other sperm. The highly proteinaceous coat can also be eaten by the soon-to-be-fertilized female, thus providing her with nutrition. But the real beauty of the spermatophore is that, among terrestrial animals, the sperm need not be introduced into the female in a liquid medium.

If you are an amphibian and lack a penis, the spermatophore comes in very handy. Although frogs, a large class of amphibia, handle penislessness by simply dumping sperm and eggs into their watery environment, many salamanders have opted for the spermatophore. One species of male salamander vaccinates the female with an aphrodisiac. As she lustfully follows him about, he lays a gelatinous spermatophore on the ground. In passing over the spermatophore, she sucks it up into her genital opening.

Among land vertebrates it is the birds that, as a group, lack a penis or any other copulatory organ—although a rudimentary one seems to have developed in ducks, geese, and swans. For the less well-endowed birds, sperm transfer is accomplished by a "cloacal kiss." The cloaca is a common opening through which the intestines, kidneys, and genital organs discharge their products. All vertebrates except mammals have cloacas. By pressing theirs together, male and female birds effect sperm transfer. Even some fish prefer the cloacal kiss to spawning and external fertilization. Who can blame them?

Most reptiles have a penis. In snakes it is even a two-headed, or bifurcated, affair—each bifurcation being called a *hemipenis*. But it is among the class Mammalia that the penis comes into its own. *All* mammals have penises, and they are usually complex organs consisting of nerve, muscle, erectile, and vascular tissues. When the erectile tissue, under proper nerve stimulation, fills with blood, the penis becomes rigid. In whales it can grow to a length of twelve feet. The world's largest testis, in case you are interested, belongs to the right whale. It is a humongous sperm factory that weighs half a ton and is more than nine feet long.

Many if not most mammals have also evolved a structure within the penis to facilitate insertion into the female genital tract. It is the *baculum* or *penis bone*. Penis bones are found in a surprisingly large number of mammals. Male bats, shrews, moles, and carnivores such as dogs, cats, bears, and weasels possess them, as do seals and sea lions. Most primates, including some of our closest relatives, have penis bones. The baculum of the walrus is a clublike two-foot-long structure.

Penis bones or pedipalps, hermaphroditism or homosexual rape, living things will find a way to perpetuate themselves sexually. There are 1½ million known species of living things in the world, and no two seem to do it in quite the same manner. That is the beauty and the wonder of nature . . . and of sex.

Earth: No Air Mask Necessary

When Neil Armstrong walked on the Moon and looked skyward on a sunlit day, he saw a pitch-black sky peppered with stars. He needed a space suit to keep from suffocating, to prevent his eardrums from exploding outward, and to protect his skin from lethal radiation—all because Earth's Moon has no atmosphere. (Actually a very thin atmosphere—about *one-trillionth* that of Earth—has been detected, but for all practical purposes, the Moon's surface is airless.)

Our Moon is, of course, not alone. Of the more than sixty moons circling the planets, only two, Saturn's Titan and Neptune's Triton, have an atmosphere. (The jury is out on Pluto's moon, Charon. Perhaps it has a very tenuous atmosphere, which, like Pluto's, is frozen into the surface most of the time.) By contrast only one planet, Mercury, does *not* have an atmosphere. Still, there's only one place where Armstrong and the rest of us can survive without the protection of a space suit: Earth. How did our planet become so well endowed?

The word *atmosphere* comes from the Greek, and it means "ball of vapor." For a planet or moon to have a ball of vapor around it today, it must, at some time in the past, have acquired that vapor and also have been able to hold on to it for more than four billion years. The formation of the solar system can help us understand the process of atmosphere acquisition.

Getting an Atmosphere

The solar system began as a vast cloud of dust and gases called the *solar nebula*. The gases were primarily hydrogen (90 percent) and helium (9 percent), the remaining 1 percent a mix of elements such as oxygen, carbon, nitrogen, argon, sulfur, silicon, magnesium, aluminum, iron, and nickel. Under the pull of its own gravity, the solar nebula coalesced. The central portion collected the greatest mass and evolved into the Sun, which reflects the original cloud composition—mainly hydrogen and helium. Other regions of the solar nebula fragmented and coalesced to become the planets and their moons. Many of the elements combined into compounds to form the solid rocky material of the inner (or terrestrial) planets and many of the moons. As for gas formation, oxygen, which is very reactive, or unstable chemically, joined with hydrogen to form water vapor. Hydrogen that was left over (after all, it *is* the most abundant element in the universe) joined with nitrogen to form ammonia gas and with carbon to form methane gas. After these processes occurred, still more hydrogen was left over. Other gases, such as helium and argon, are very nonreactive, or stable chemically, and did not react at all with the other elements.

The equations in Figure 1 summarize the flurry of activity that took place among the gases early in the history of the planets and their moons.

The solar system at this point was still in its infancy. The material of the planets and their moons was still contracting. Much of the mixture of gases that was a part of the original gas cloud or that formed from the subsequent chemical reactions got trapped in the spaces of the solid rocky component of these bodies. But as gravity continued

Figure 1

(1) hydrogen + oxygen $\longrightarrow$ water vapor
$2H_{2(g)}$ $O_{2(g)}$ $2H_2O_{(g)}$

(2) hydrogen + nitrogen $\longrightarrow$ ammonia
$3H_{2(g)}$ $N_{2(g)}$ $2NH_{3(g)}$

(3) hydrogen + carbon $\longrightarrow$ methane
$2H_{2(g)}$ $C_{(s)}$ $CH_{4(g)}$

its relentless squeeze, these gases percolated out, much as water is squeezed out of a sponge. They surrounded the solid portion of their moon or planet, blanketing it in a gaseous envelope—an atmosphere. This atmosphere was largely hydrogen, helium, ammonia, methane, and water vapor. (Yes, Earth in its early evolution had this type of atmosphere.) What caused Mercury and most of the moons to lose their atmosphere?

Keeping an Atmosphere

It is one thing to form an atmosphere and quite another to hold on to it. Whether or not a planet or moon can hold on to the molecules that make up its atmosphere and keep them from flying off into space depends on several factors.

Mass and Size of a Body

Mass and size combine to give a body its *surface gravity*, or the gravitational attraction it has at its surface. The gravity of a planet or moon attracts a molecule of gas, invisibly holding on to it, preventing it from escaping. On Earth, a molecule of gas would have to attain a speed or velocity of 7 mi/s (11.2 km/s) to overcome its gravitational attraction and escape into space. This is known as the body's *escape velocity*, and it is dependent on its surface gravity.

Generally, the more massive *and* smaller a body is, the greater are its surface gravity and escape velocity. This is because surface gravity is a function of a body's mass *and* the distance from its center to its surface. If two planets are the same size, the more massive one will have the greater surface gravity and escape velocity. If they have the same mass, the smaller one will have the greater surface gravity and escape velocity. In fact, if Earth were squeezed to the size of a marble (same mass, smaller size), its escape velocity would increase to beyond the speed of light. Earth would become a black hole.

Table 1 gives a list of the escape velocities of our solar system's nine planets and three of its large moons (in order of decreasing value):

Table 1

Planet/Moon	Escape Velocity— mi/s (km/s)
Jupiter	37.0 (59.6)
Saturn	22.1 (35.6)
Neptune	15.3 (24.6)
Uranus	13.1 (21.1)
Earth	7.0 (11.2)
Venus	6.5 (10.4)
Mars	3.1 (5.0)
Mercury	2.7 (4.3)
Titan	1.5 (2.5)
Moon (Earth's)	1.5 (2.4)
Pluto	Low
Triton	Low

Judging by Table 1, Jupiter, Saturn, Neptune, and Uranus should have extensive atmospheres, and indeed they do.

But how about Titan? It has an escape velocity lower than Mercury and about the same as Earth's Moon, yet it has an atmosphere, whereas Mercury and Earth's Moon do not. In fact its atmospheric pressure is *1½ times* as great as Earth's, though Earth's escape velocity is nearly *five times* as high. Clearly, escape velocity, or mass and size of a body, is not all there is to consider in atmosphere retention.

Temperature

Temperature also plays a critical role. It is a measure of the energy of motion of a group of molecules. If the temperature is high, the molecules are moving faster; if it is low, they are moving slower. (See "How Hot Is Hot? How Cold Is Cold?") It makes sense, therefore, that a planet or moon with a higher temperature will have in its atmosphere more gas molecules that are able to reach escape velocity and fly away. Temperature is the primary reason why Titan has a considerable atmosphere despite its small surface gravity and escape velocity. It is far away from the Sun and quite cold. The molecules in its atmosphere move slowly and can be held on to rather easily. Pluto and Triton, which have even smaller surface gravities and escape velocities, have atmospheres for the same reason. Mercury and Earth's Moon, on the other hand, are much closer to the Sun and much hotter. The molecules that once made up their atmospheres moved too rapidly to be held. Most other moons do not have an atmosphere simply because their mass and surface gravity are too low to retain one.

Mass of Gas Molecules

The types of gases that make up an atmosphere also play an important role in their retention. Imagine a section of atmosphere with gas molecules bouncing around randomly, like billiard balls on a pool table. As they collide, their energy of motion is transferred among one another. The lighter molecules—the lightest gas molecules being hydrogen and helium—gain more velocity than the heavier ones when hit. It is therefore more likely that they will reach escape velocity during any given collision. If such molecules are at the outer edge of the atmosphere, where they are unlikely to collide with other molecules, they may escape into space.

To summarize, the ability of a planet or moon to retain an atmosphere depends on (1) the body's holding power, or gravitational attraction—more massive bodies are generally better at holding an atmosphere; (2) the temperature—a body can hold an atmosphere more easily at colder temperatures; (3) the types of molecules that make up the atmosphere—the body can hold on to heavier or more massive molecules more easily than lighter ones.

A phenomenon called the *solar wind* is also thought to play a part in atmosphere retention or loss. The solar wind is an outflow from the Sun of high-speed, charged particles. These particles have a tendency to "blow off" an atmosphere. The closer a planet or moon is to the Sun, the more it is affected by the solar wind.

Earth provides a classic illustration of all of this. As mentioned earlier, Earth in its infancy had an atmosphere made up of hydrogen, helium, ammonia, methane, and water vapor. Of these gases, hydrogen and helium were too light to be held by Earth's gravitation. Given Earth's temperature, they were able to reach escape velocity and leak away into space. The remaining molecules were large enough and moved slowly enough to be retained. Earth's first permanent atmosphere was established. It consisted of ammonia, methane, and water vapor and is referred to as *Atmosphere-I, or A-I*. Chemists call such an atmosphere, which contains molecules rich in hydrogen (NH_3, CH_4, H_2O), a *reducing atmosphere*.

Early Evolution of Earth's Atmosphere

The combination of reducing gases that comprised A-I of early Earth, and probably of Venus and Mars as well, was stable enough. By themselves the gases would not react and undergo further chemical change. But nature provided a push in the form of outside energy. Heat from volcanic activity, electricity from lightning discharges, and, most important, high-energy radiation from the Sun kicked these gas molecules up to higher energy states, making them unstable. A particularly key player in this scenario was ultraviolet radiation. UV acts like a little hatchet on molecules, and it was especially adept at splitting water into its elemental parts, hydrogen and oxygen:

$$2H_2O_{(g)} \xrightarrow{UV} 2H_{2(g)} + O_{2(g)}$$

The process is known as *photolysis*, which comes from the Greek words *photo*, meaning "light," and *lysis*, meaning "to split."

The gaseous hydrogen molecules, being extremely light and nimble, moved off into space and were lost. Oxygen was heavier and more lumbering and hung around. Its presence in the atmosphere changed things dramatically. Although ammonia and methane are stable in the presence of each other and of water, they are decidedly *unstable* in the presence of oxygen. What happened was quite interesting. The oxygen molecules combined (1) with the carbon and hydrogen of methane, forming carbon dioxide and water, and (2) with the hydrogen of ammonia to form water, leaving nitrogen gas as a by-product. Figure 2 depicts these chemical processes.

Figure 2

(1) methane + oxygen ⟶ carbon dioxide + water vapor
    $CH_{4(g)}$ $2O_{2(g)}$ $CO_{2(g)}$ $2H_2O_{(g)}$

(2) ammonia + oxygen ⟶ nitrogen + water vapor
    $4NH_{3(g)}$ $3O_{2(g)}$ $2N_{2(g)}$ $6H_2O_{(g)}$

What effect did this bustle of activity have on the composition of the atmosphere? Very simply, a methane/ammonia/water vapor atmosphere changed to a carbon dioxide/nitrogen/water vapor atmosphere. Or A-I changed to A-II.

(Volcanic activity may also have played a role in adding carbon dioxide and water vapor to the atmosphere. The impact of volcanism on the transition from an A-I to an A-II atmosphere is uncertain.)

How long it took for this change to happen is difficult to assess, but most certainly by the year four billion B.C. (when Earth was about half a billion years old), A-I had become largely A-II. Venus today retains an atmosphere that is principally carbon dioxide, nitrogen, and water vapor. Mars, too, is mainly carbon dioxide and nitrogen. Earth, however, diverged from its neighbors and was in for yet more dramatic changes.

Making Earth's Atmosphere Breathable

Let's do a bit of time travel. Put on your gas masks, for we are traveling back to Earth as it existed several billion years ago. The atmosphere is A-II. Relentless hammering of the Sun's UV rays continues to split water molecules into hydrogen and oxygen. (Remember photolysis?) Hydrogen continues to fly off into space, much as a balloon filled with hydrogen gas will do. Oxygen remains to bother other molecules, including solid parts of Earth's crust. But a point has been reached where there are no longer other molecules to bother. All of the methane has been converted to carbon dioxide and water and the ammonia to nitrogen and water. Silicon and iron in the crust have combined to form oxides. What's an oxygen molecule to do?

Triple up, that's what! Normal oxygen gas exists as two-atom molecules: O_2. It's the oxygen we breathe. But under the action of the Sun's high-energy UV radiation, normal oxygen can be made to join as three-atom molecules (see Figure 3).

This O_3 gas is called *ozone*, and it is not breathable. However, it has another very important property that ordinary oxygen does not have: it blocks out most of the UV and other high-energy radiation produced by the Sun.

Figure 3

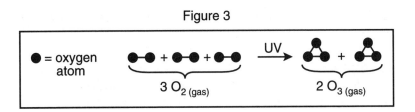

In itself this does not seem like much. But as the following essay will explain, without the ozone layer, human life on Earth never would have evolved.

In any event, a layer of ozone formed as a result of the action of UV on O_2 in the upper atmosphere. This layer in turn blocked the further penetration of UV, so photolysis of water molecules came to a screeching halt. Generation of oxygen ceased, and the A-II atmosphere stabilized, with very breathable oxygen. This could have been the end of the story.

It was not, however, because some three billion years ago a truly miraculous development occurred on Earth—and nowhere else in the solar system, or in the universe as far as we know. The miracle, in a word, was *chlorophyll*. Through a series of fortuitous accidents called evolution, living things developed the chemical machinery to put together chlorophyll molecules. With the construction of chlorophyll, water could once again be split into hydrogen and oxygen, because UV was not necessary. Visible light could do the trick. Such is the wonder of chlorophyll. The process was called *photosynthesis*, and over millions of years it added O_2 to our atmosphere, making it breathable. The hydrogen gas that separated from the water did not escape but was made to attach to carbon dioxide, forming food molecules.

As a result of all this photosynthesis, the A-II composition of carbon dioxide/nitrogen/water vapor changed to one of nitrogen/oxygen/water vapor. This is the atmosphere we enjoy today: 77 to 78 percent N_2 and 21 percent O_2 (in dry air), with variable amounts of water vapor (up to about 4 percent) and traces of other gases, including argon and carbon dioxide. It is called *A-III*, and it is rich in free, uncombined, ordinary oxygen, O_2. Hence chemists refer to A-III as an *oxidizing atmosphere*.

It has taken the better part of $4\frac{1}{2}$ billion years for our atmosphere to form and then to *trans*form from a suffocating blend of cooking gas, ammonia, and explosive hydrogen to the life-sustaining mix we inhale today. It is the only place in the cosmos we know of that has a breathable atmosphere. The chemical balance that keeps it so is a delicate one. Too much CO_2 in the atmosphere from the burning of fossil fuels or a significant reduction of the ozone layer can upset this balance and turn a Garden of Eden into the hellish furnace that is Venus or a frigid world bathed in lethal radiation, such as Mars. Please, let us not allow that to happen.

How Did Life on Earth Start?

Life exists almost everywhere on Earth—deep in the oceans, on mountaintops, and even higher (bacteria have been found at 25½ miles above sea level, nearly five times higher than the highest mountain)—in the hottest and the coldest regions of the world. Wherever you go, there is life. If you doubt this, try to keep weeds out of your garden or food from becoming moldy.

Has this always been true? We seem to find life no matter how far back we go: the oldest rocks—close to four billion years old—contain fossil records of once-living things. So it is natural to wonder whether life on Earth predates Earth. Did it originate on Earth or elsewhere, in some dark and distant corner of the universe?

Extraterrestrial Origins

According to one school of thought life on Earth indeed predates Earth—it began elsewhere and has been carried through space to "seed" planets with suitable conditions. The concept is called *panspermia*, and goes back thousands

of years. The ancients envisioned Earth as a fertile garden
seeded with the spores of extraterrestrial life. In the nineteenth century distinguished scientists, including Lord Kelvin (see "How Hot Is Hot? How Cold Is Cold?"), suggested that dormant forms of life had been transported to
Earth by meteorites. The theory is known as *lithopansper-
mia*, from the Greek word for stone—*lithos*. As recently as
1909 the great Swedish scientist Svante Arrhenius stated
that microorganisms might be driven from place to place in
the universe—including Earth—by radiation pressure, a
theory known as *radiopanspermia*.

Over the greater part of the twentieth century the concept of panspermia has lost favor within the scientific community. Repeated attempts to detect life activity in meteorite
fragments have failed. Also, most scientists feel that spores
or seeds drifting through the universe, as durable as they
might be, would not be able to survive the temperature and
radiation rigors of deep space. Nonetheless, the belief that
life originated "out there" still has its proponents—among
them the renowned English astronomer Fred Hoyle.

Although panspermia is no longer generally accepted,
many astronomers—such as Carl Sagan of Cornell University, who has called them "the bringers of life"—believe that
asteroids, comets, and meteorites bombarding Earth, as well
as interplanetary dust "raining" down, *did* provide important seeding, not of life itself but of organic compounds
that were critical precursors in its evolution. According to
Armand H. Delsemme of the University of Toledo, meteors
could have brought water, without which there can be no
life (see "The Miracle Molecule"), as well as important
organics. These are just two of the players in a growing
controversy among scientists as to the exact role celestial
bodies played in the origin of life on Earth. Others, such as
Stanley Miller of the University of California at San Diego,
believe that comets, asteroids, and the like played an incon-
sequential role in life's beginnings.

Terrestrial Origins

Whatever the role of asteroids in seeding Earth with the raw materials for life, the consensus view among scientists is that the origin of life was essentially an Earthbound occurrence, so let us proceed on that premise. Several possibilities exist. Life may have begun because of a supernatural event or through natural chemical and physical processes. Supernaturalism is religious rather than scientific in nature and not supported by experimentation and observation. It does not belong in a book of science exploration. That leaves the origin of life through natural processes. Searching for the way to create life has occupied necromancers and novelists through the ages. Mary Shelley's Dr. Frankenstein may have succeeded in reviving dead tissue by harnessing the electric power in lightning, but in the real world life was generated from nonlife by nature alone, billions of years ago, when very specific conditions were just right.

Spontaneous Generation

The ancient philosophers believed that mud, warmed by the Sun, could transform itself into frogs and snakes, that old rags could turn into mice and rats, that spoiling meat could become maggots. As late as 1609 a French botanist wrote, "There is a tree. . . . From this tree leaves are falling; upon one side they strike the water and slowly turn into fish, upon the other they strike the land and turn into birds." Other writings allude to people arising spontaneously from sperm injected into cucumbers and microorganisms arising spontaneously from broth or gravy. The belief that living things can arise at any time from nonliving matter by some kind of metamorphosis is known as *abiogenesis*, or *spontaneous generation*.

The concept certainly provides an easy explanation for the origin of life. The only problem is that under rigorous scientific analysis it did not hold up. An Italian physician, Francesco Redi, in a classic set of experiments conducted in 1668, proved that maggots could not arise from rotting meat. A century later Lazzaro Spallanzani, another Italian investigator, demonstrated that boiled broth would not give rise to microorganisms. Unfortunately his findings were questioned by other investigators, and the debate over spontaneous generation went on for another century. In the mid-1800s two scientists working independently, Louis Pasteur in France and John Tyndall in England, proved conclusively that living things could not arise from broth—or from any other nonliving matter. The concept of spontaneous generation had sounded its death rattle.

With the discarding of abiogenesis a paradox of sorts was created. If life can come only from life, and living things did not exist on Earth at the time of its formation (of which scientists are reasonably certain), then where on Earth (no pun intended) did the *first* living things come from?

Prebiotic Evolution

At some time in the distant past life had to have come about from nonliving matter. Do not confuse this notion, however, with that of spontaneous generation. The creation of life was a gradual and complex process, taking many millions of years. It required a constant source of high energy that is no longer available. It required an atmospheric composition that no longer exists. The origin of life, though probably not a one-shot deal (it occurred repeatedly over different portions of the primordial Earth), *did not* happen easily and in all likelihood has not offered an encore performance within the last few billion years.

The tedious process of matter increasing in complexity from inorganic to organic to living is referred to as *chemical* or *prebiotic evolution*—evolution before life. It has also been called the *chemosynthesis hypothesis* and was first proposed in the 1920s by English biologist J. B. S. Haldane and Russian biologist Alexander I. Oparin.

Haldane and Oparin's theory is based on our understanding of the Earth's nascent atmosphere four billion years ago (see "Earth: No Air Mask Necessary" for details). Made up largely of hydrogen, ammonia, methane, and water vapor, the atmosphere later lost hydrogen, and chemical processes converted the ammonia and methane to nitrogen and carbon dioxide. High-energy solar radiation—ultraviolet (UV) in particular—continually bombarded this atmosphere. Lightning was common. Earth was hot from gravitational contraction and from the decay of radioactive elements. With this mix of matter and energy, scientists feel, conditions on Earth were suitable for the evolution of life.

But did it happen? Ever since the chemosynthesis hypothesis was first proposed, scientists have been testing it. In 1953 American biologist Stanley Miller subjected a "primordial atmosphere" of methane, hydrogen, ammonia, and water vapor to electrical sparks and boiling. After one week amino acids and sugars were found in the reaction chamber. Amino acids are essential constituents of all living things— they are the building blocks of proteins. Sugars are important as nutrients and as structural components of cells.

Sidney Fox, of the University of Miami, tested the next step in the chemosynthesis hypothesis. In 1957 he heated a mix of eighteen different amino acids and got large, protein-like molecules. Other investigators were successful in synthesizing molecules called *nucleotides*, which are the building blocks of DNA and RNA—the genetic material! American physical chemist Philip Abelson in 1956 tested the hypothesis using the more modern atmosphere, which contained nitrogen and carbon dioxide instead of ammonia and methane (just *in case* life formed in this environment) and got amino acids and other organics as well.

Curiously, when oxygen gas—O_2—was added to the mix in the reaction chamber, no organic synthesis occurred.

Oxygen is very reactive chemically, combining with elements and compounds around it to form very stable, or nonreactive, products. Molecules must be relatively *unstable* to combine and form more complex molecules on the way to becoming alive.

Although no experiment created life, even in its simplest form, the success of these and other experiments gave credibility to the hypothesis of chemical evolution, which scientists believe proceeded this way: In the warm oceans of an Earth less than half a billion years old, organic molecules accumulated. Amino acids formed and joined into proteins; fats and oils formed; sugars formed and combined into starches; nucleic acids (DNA and RNA) formed from nucleotides. All of the essential ingredients for life were present as the oceans of Earth became a rich organic soup. Liquid water provided an ideal medium in which molecules could mix and react. The jump from organic hodgepodge to life, in the form of primitive, membrane-bound, cell-like structures, with a serviceable enzyme system to speed up vital chemical reactions and a crude but functional ability to reproduce, took perhaps several hundred million years.

Life had begun. The evolution of this life from the simplest of cells to the staggeringly complex and diverse forms we see today took another $3\frac{1}{2}$ to 4 billion years.

Biotic Evolution

The first primitive cells lacked a nucleus and many other subcellular structures, or organelles, present in more advanced cells. They were called *prokaryotes*, and they exist to this day, in the form of bacteria. These cells were unable to make their own food; instead they sponged off the organic molecules that were dissolved in the oceans in which they lived. This method of nutrition—the taking in of food already made—is called *heterotrophism*. The cells were also unable to use oxygen to break down food molecules for energy. The process of releasing energy from food is called *respiration*; without oxygen it is called *anaerobic respiration*. The first living beings were indeed anaerobic. This makes sense, since free, or uncombined, oxygen was not yet present in the atmosphere.

The problem with anaerobic respiration is that it is terribly inefficient. It requires a lot of food to produce very little energy. Alas, as organically rich and soupy as the oceans were, the food they contained would not last forever. Living things were using up that food a lot faster than chemical processes were replenishing it. Without food life cannot exist. Would the fate of life be to starve itself into extinction long before even the lowliest worms crawled onto the land?

But for a magic molecule called *chlorophyll*, the answer might be yes. Chlorophyll evolved some $2\frac{1}{2}$ to 3 billion years ago and dramatically changed the course of evolution. It provided a method of capturing the energy in sunlight and converting it into stored chemical energy, or food. Thus living things were no longer dependent on the depleting reservoir of nutrients built up through prebiotic evolution. They could make their own food by simply basking in the Sun. Nutrition in which living things make their own food from basic raw materials is called *autotrophism*. Using sunlight to get the job done is called *photosynthesis*. By the way, photosynthesis also released free oxygen into the atmosphere (see "Earth: No Air Mask Necessary"). Chemical analysis of rocks suggests that our atmosphere became *aerobic* about two billion years ago.

With free oxygen present, aerobic life soon evolved. Using oxygen, cells could extract *nearly twenty times* as much energy from nutrients as they could without oxygen. Food supplies lasted longer, and more elaborate organisms evolved. Simple prokaryotic cells became more organized, giving rise to the first *eukaryotic* cells, complete with their own membrane-bound nucleus. This happened about 1.4 billion years ago. Sometime shortly after, eukaryotic cells began to assemble into multicellular organisms. The oldest fossils of multicellular animals appear about one billion years ago. All of today's animals and plants are comprised of eukaryotic cells.

Life also wriggled onto the land sometime after oxygen become part of the atmosphere. In fact it is likely that land-based life would never have evolved were it not for free oxygen. The reasoning goes as follows: Life evolved in the seas through chemosynthesis. It was a delicate process. The newly formed organic molecules were highly structured and fragile. The intense solar radiation that assaulted Earth, especially in the form of UV, could easily dismantle them as quickly as they formed. It is unlikely that even the simplest cells would have had a chance to evolve under these conditions. If, however, as these molecules formed they sank below the surface of the water, perhaps several inches or a foot, they would be protected largely from the molecule-rending radiation. Chemosynthesis could proceed without a hitch. But on land life would be battered constantly by UV.

Enter oxygen. As discussed in "Earth: No Air Mask Necessary," UV and lightning discharges converted ordinary oxygen in the upper atmosphere to ozone (see Figure 1).

Figure 1

$$\text{ordinary oxygen} \xrightarrow{\text{radiation or lightning}} \text{ozone}$$
$$3O_2 \qquad\qquad\qquad 2O_3$$

Ozone has the unique ability to absorb UV. Thus, with the establishment of a shielding layer of ozone (which still exists today, though environmentalists are concerned that it is being depleted), life could—and did—move onto the land, where it evolved over some billion years into slugs, bugs, frogs, snakes, birds, ferns, flowers, cucumbers, . . . and people. Without a protecting ozone layer, such evolution would have been highly unlikely.

As far as we know, life did not evolve anywhere else in the solar system. Direct testing of Martian and lunar soil have shown no life activity. Venus is far too hot. Mercury is alternately too hot and too cold and has too much radiation. The gas giants—Jupiter, Saturn, Uranus, and Neptune—are not made of the right stuff and are too cold at the surface. Pluto is a frozen wasteland. Perhaps some of the more than sixty moons have simple, microscopic life on them, but don't bet on it. The evolution of life is a precious thing requiring the right kinds of matter, the right kinds of energy, the right temperature, liquid water, and time. For this life to become complex and diverse, conditions must change at critical points so that heterotrophic life can give rise to autotrophic life, anaerobes can give rise to aerobes, and aquatic forms can slither out of the oceans onto the land. So intelligent life can evolve and, in turn, wonder about its origins.

As far as we know, these miracles happened on Earth and nowhere else.

Tropical Rain Forests:
It's a Jungle Out There

"The land is one great, wild, untidy luxuriant hot-house made by Nature for herself." This is how Charles Darwin described the tropical rain forests when he first beheld them more than a century and a half ago. If anything, he understated what is a maddening celebration, a riotous dance of life. To explore the exquisite beauty of the tropical rain forests, with their nearly infinite variety of sights and sounds and smells, is to visit an unknown, magical planet.

Almost all tropical rain forests lie on either side of the equator, between the tropic of Cancer to the north and the tropic of Capricorn to the south. It is an equatorial belt that circles our planet like a giant cummerbund around a fat belly. Within that belt, lush rain forests abound, each forest its own garden of Eden. They are the jungles of the world. South America boasts the largest of these Edens, the vast *Amazon Hylaea* of Brazil and several neighboring countries. Covering an area nearly the size of Australia, it is named after the greatest of all rivers, which snakes through it. So large and powerful is the four-thousand-mile Amazon River that where its mouth empties into the Atlantic Ocean you can travel one hundred miles out to sea and scoop up a glass

of fresh water. One-sixth of the fresh water flowing over
Earth moves through the Amazon River's vast drainage
system.

Other rain forests flourish in Central America, South-
east Asia, Malaysia, Indonesia, and a small section of central
Africa (mainly in Zaire). Even the United States is represented by small rain forests in Hawaii, Puerto Rico, and the
Virgin Islands. Figure 1 shows the distribution of the
world's major tropical rain forests.

Figure 1

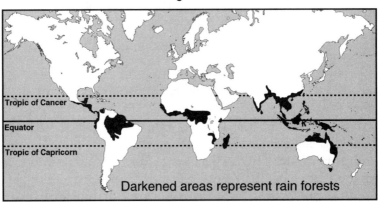

Darkened areas represent rain forests

Though most are clustered around the equator, not all
rain forests are alike. Scientists recognize more than forty
different types, each with its own distinctive flora and
fauna. More broadly, tropical rain forests can be divided
into three categories: *tropical dry forests*, *tropical moist forests*, and *true tropical rain forests*. True tropical rain forests—the most common of the three and the ones generally
discussed here—are found nearest the equator and at low
altitudes. They are the wettest land areas in the world, being
drenched by up to four hundred inches of rain annually, as
compared to forty-three inches in New York City and twenty
inches in San Francisco. Temperatures in the tropical rain
forests average 80°F, seldom varying much from the mean.

Life thrives in all of this heat and humidity, and it has

made tropical rain forests the most complex and biologically diverse environments on the planet. They comprise roughly 7 percent of the total land surface of Earth, yet they are home to perhaps 90 percent of the planet's animal and plant species. Rain forests have been called, and rightly so, "the most alive places on Earth."

Biodiversity and the Rain Forest

While the richest forests in the continental United States can boast 25 different tree species at most, in the rain forests of Borneo alone twenty-five hundred arboreal species have been identified. Amazingly, packed into a two-and-a-half-acre plot of Brazilian Amazon were 450 different species of trees. The tallest of these rain forest trees tower over two hundred feet above the forest floor. They are the *emergents*, scattered giants whose leafy heads poke out above the dense foliage of the *canopy* below.

The canopy of the rain forest is a dense tangle of treetops some 100 to 120 feet above ground level. Roughly two-thirds of all the flora and fauna of the rain forest live or dwell in the canopy, and it is here that nature puts on its most spectacular display. Draping the canopy like a magnificent aerial botanic garden are twenty thousand different species of *epiphytes*—plants that grow on trees but do not parasitize or harm them. The roots of epiphytes dangle freely, grabbing water and nutrients from the air or from debris collecting in the many crevices of the tree's trunk and branches.

Orchids are by far the best known and most beautiful epiphytes. Most of the world's orchids are found in tropical rain forests and nowhere else. Here, over millions of years, they have coevolved with animals in often remarkable ways. One species of orchid, for example, has a flower that mimics the general body design of a particular type of female insect. Pollination is achieved when the hapless male is duped into having sex with the orchid.

Bromeliads are another type of epiphyte. Close relatives

of the pineapple, they are designed to catch rainwater in the base of their leaves. This standing water creates a minieco-system within which insects and even tiny amphibians such as the poison dart frog (also called the *arrow poison frog*) can thrive. The poison dart frog is brightly colored and highly poisonous, providing forest hunters with a powerful toxin for their arrows and blowgun darts. Voodoo practitioners find the poison useful for transforming people into "zombies."

The canopy of the rain forest is also abuzz with a wild assortment of flying critters. Half of all the world's birds—that's several thousand species—live in the Amazon rain forest alone. Tiny hummingbirds, with their enormously elongated bills, are important pollinators, flitting tirelessly from one flower to another. Their wing speed of fifty to eighty beats per second is so fast that, were a human to expend such energy, he would have to consume the caloric equivalent of 130 loaves of bread daily.

Brilliantly colored parrots, macaws, toucans, and birds of paradise adorn the green canopy like so many priceless jewels. Their loud and raucous screeching is a familiar sound in the rain forest. But it is the hawk and the eagle that rule from their canopy perches, swooping down and killing any animal not swift enough to avoid their attack. They are at the top of their particular rain forest food chain.

Perhaps most extraordinary are the huge numbers of bats that live in the tropical rain forests. They are the only true flying mammals, and most of the nearly one thousand known bat species are indigenous to rain forests. In some areas more than half of all mammal species are bats. Their biomass can be greater than that of all other mammals combined.

Bats range in size from tiny insect eaters to huge fruit-eating "flying foxes," with a wingspan of up to six feet. Three species are vampires, using razor-sharp triangular front teeth to slice into victims. An anesthetic in the bat's saliva makes the procedure painless, while an anticoagulant keeps the blood flowing as the bat laps at its sole source of nourishment.

Like insects and birds, bats are important pollinators of rain forest plants. It is, however, as seed dispersers that they perform their greatest service. By eating fruits and then flying off, they spread seeds throughout the tropical rain forest. It has been estimated that as much as 95 percent of the seed dispersal in a rain forest is accomplished by bats.

Beneath the canopy smaller trees, shrubs, and vines compete for space and for the very limited amount of light that filters through the thick overhanging vegetation. These lesser flora form the *understory*, a tangle of leafy branches fifty to eighty feet above the forest floor.

Like the canopy, the understory is teeming with life. Swinging or climbing from one level to another are most of the world's primates—monkeys and apes. In the jungles of Central and South America, woolly monkeys, spider monkeys, and the shrill howler monkeys abound. Common to them all is the New World feature of a *prehensile tail*, one that can grasp tree limbs. It has evolved as a fifth limb to aid in an arboreal way of life. No Old World (Asian, African) monkeys have this type of tail.

Chimpanzees are found throughout the African rain forests. Being apes (along with gorillas, gibbons, and orangutans) and not monkeys, they have no tails and feel as much at home foraging for food on the forest floor as they do scampering about the canopy. The gorilla, largest of the apes, is also a creature of the African rain forest. Due to its great size, the adult gorilla rarely if ever climbs a tree.

An unusual tree dweller and one of the most common large mammals of Central and South American rain forests—making up two-thirds of the weight of all living mammals in some areas—is the sloth. It is unusual because it is almost always seen hanging upside down from tree branches, its long, curved claws acting like meat hooks. Sloths eat, sleep, mate, and even give birth in this position. Algae usually form a thick infestation in their fur, giving sloths a greenish color, which actually serves them well as a protective camouflage. Unlike other mammals, a sloth's body temperature varies with that of the environment. For this reason sloths enjoy sunning themselves (hanging upside down of course) wherever an opening in the canopy can be found.

Prowling the Amazon Hylaea is perhaps the true king of the South American jungles—the jaguar. Third largest of all cats, behind the lion and tiger, it rules from forest floor to canopy, pouncing on any weaker, slower animal. Atop its food chain, it has no natural predators.

In summary, no ecological niche goes unfilled in the rain forest. This has created some highly unusual creatures to say the least. Arnold Newman describes some of them in *Tropical Rainforest*:

> The parade of species reads like a catalog of the inconceivable. Grass in the form of bamboo grows 100 feet high, at a rate of 36 inches in 24 hours. There are "roses" with 145-foot trunks; daisies and violets as big as apple trees; 60-foot tree ferns with some of the hardest wood to be found; 37.5-foot constricting snakes and lily pads over five feet in diameter which can support a child's weight. . . . Rafflesia boasts the world's largest flower—38 inches across weighing 38 pounds and holding several gallons of liquid in its nectaries. Here too are . . . moths with 12-inch wingspans; frogs so big they eat rats, and rodents themselves weighing over 100 pounds.

Unusual to be sure. Unusual and often dangerous. In the backwaters of the mighty Amazon River and its many tributaries lurks the ubiquitous American crocodile. Schools of piranha, their razor-sharp teeth bared, have stripped two-hundred-pound animals to the bone in minutes. But most diabolical of all river inhabitants is the dreaded candiru, a spiny fish that may reach two inches in length. Candiru are attracted by the smell of urine and will enter the urethral opening of the genitalia. They can make their way up the urethra and lodge in the bladder, where surgical removal may be necessary. Needless to say, the pain they inflict can be excruciating. Males of many Amazon tribes often tie off the foreskins over the urethral openings of their penises to prevent invasion of the candiru.

Even the lowly ant attains grander dimensions in the tropical rain forests. Most notable are the driver ants of Africa and the army ants of South America. Hordes of several hundred thousand or more of these insects, each an inch in length, march through the jungle destroying everything in their path. Anything that is unable to escape—a horse, a cow—is eaten alive.

With driver ants and piranha and highly venomous tree snakes slithering about the canopy, it might come as something of a surprise that none of these are the most harmful animal in the tropical rain forest. It is the tiny, insignificant mosquito, whose bite (that of the female) can transmit the protozoan causing malaria—still probably the number-one disease killer worldwide. Mosquitoes also spread the microscopic filarial worm, responsible for elephantiasis. This disease derives its name from the immense swelling the worm causes, often turning a normal leg into one resembling an elephant's. One man, whose genitalia were affected, had to carry his testes in a wheelbarrow when he walked.

Destruction of the Rain Forests— A Worldwide Catastrophe

According to figures in *The Rainforest Book*, by Lewis Scott, our rain forests are systematically being destroyed to the tune of 67 acres a minute—a football field a second. Each year an area the size of New York State vanishes forever. "If this pace continues," Scott warns, "most of the rain forests will be gone by the end of the century."

And with the disappearance of the rain forests, we are

losing—forever—a vast number of species. They are becoming extinct. Many are endangered right now. The orange-furred orangutan (meaning literally "old man of the forest" because of its humanlike face) used to roam all over Asia. Now, due to the destruction of its habitat, the only true arboreal great ape is found, in ever decreasing numbers, only on the Indonesian islands of Sumatra and Borneo.

Just how many other animals and plants have become extinct or are on the verge of extinction? Nobody knows. Nobody even has a clue. How could they, when as many as 95 percent of the planet's living things may not have been discovered yet? So far about 1.5 million different species of organisms have been identified, half of them insects. But the total may be 10 million or even as high as 100 million. And most live in the tropical rain forests.

In *Tropical Rainforests*, Newman makes an interesting observation: We are able to measure the distance from Earth to the Moon with pinpoint accuracy (a tolerance of less than a quarter of an inch), but we can only identify less than 6 percent of the species that inhabit our planet. I guess if you don't know what's out there, you don't mind killing it. But we know enough to see that we do have much to lose if we allow the rain forests to disappear:

- One-quarter of all medicines owe their origin to rain forest plants and animals. The U.S. National Cancer Institute has identified three thousand plants as having anticancer properties. Seventy percent grow in the rain forests. The rosy periwinkle has given us several drugs that work wonders against childhood leukemia and Hodgkin's disease. Quinine, the wonder drug for treating malaria, comes from the bark of the tropical cinchona tree. Squibb Pharmaceuticals uses the venom of a Brazilian pit viper to manufacture a drug for high blood pressure.
- At least twenty-five hundred potential new fruits and vegetables are growing in the world's rain

forests. And genes from rain forest varieties of domestic plants and animals are being used to improve the homegrown stock. All domestic chickens are descendants of four species of Asian jungle fowl. Coffee, chocolate, in fact fully 80 percent of everything we eat derives from the rain forest.

• Rain forests provide us with rubber for our tires and gums and oils for our chewing gum, paint, and cosmetics. From one particular rain forest tree comes the tough latex used for golf ball covers. There is even a tree in Brazil, the copaiba, which pours out diesel fuel when it is tapped. Twenty percent of Brazil's diesel fuel is now being supplied by this tree. Its discoverer, Melvin Calvin, won a Nobel Prize for his efforts.

So why destroy the source of such valuable resources? There are several reasons for the destruction of the tropical rain forests. Logging companies fell huge numbers of trees and sell the hardwood lumber internationally. Japan, incidentally, is the world's largest importer of tropical lumber, using enough wood in disposable chopsticks in one year alone to build eleven thousand single-dwelling homes.

Trees are also chopped down to provide land for cattle ranching, which has become big business in the rain forests. Unfortunately, the endeavor is highly inefficient. It has been estimated that sixty-seven square feet of Brazilian rain forest must be sacrificed to produce a quarter-pound hamburger. The soil in the rain forest, contrary to popular belief, is very thin and fragile, and the constant torrential rains wash away whatever topsoil exists. Within several years ranchers must abandon their grazing lands and search for newly deforested areas.

In an operation termed *slash and burn*, much of the tropical rain forest is being cleared to provide land for farming. Trees are felled and then set afire, the ashes adding much-needed nutrients to the soil. Agriculture nonetheless suffers the same fate as cattle ranching. Topsoil is quickly eroded away, and the farms soon must be abandoned. A sizable chunk of forest has been cleared to grow coca for the illicit drug trade. In 1988 cocaine was Colombia's number-one export, grossing $4 billion. Coffee, at $1.5 billion, was a distant second.

When we chop down and torch rain forest trees, we are doing much more than wiping out untold numbers of plant and animal species. As we destroy the planet's diversity of life, we may also be ruining the atmosphere on which all life depends. This is how it happens.

Green plants (and algae) are special kinds of living things. They take carbon dioxide from the air and magically combine it with water sucked up from the soil (or sea), forming a simple sugar, glucose, and oxygen gas. This magic is performed by the planet's green pigment, chlorophyll, which captures sunlight and uses it to power the chemical reactions collectively termed *photosynthesis.*

By taking carbon dioxide from the air and locking it into glucose, trees effectively lower the atmospheric levels of carbon dioxide. Tropical rain forests, with their abundant growth of trees, are a major player in this photosynthesis game. Annually, they remove millions of tons of carbon dioxide from the air.

Reducing atmospheric concentrations of carbon dioxide is a good thing, for carbon dioxide is a greenhouse gas, and we are already producing too much of it from industrial and automobile pollution. Greenhouse gases absorb heat that is radiating from Earth's surface, thereby preventing its escape. As a result things heat up and we experience "global warming."

Enter the slash-and-burn farmer. Slashing and burning is actually a double-edged sword. First trees are cut down, an activity that reduces photosynthesis and ultimately increases levels of carbon dioxide in the atmosphere. (It will also decrease the levels of atmospheric oxygen gas, a necessity of life. Photosynthesis is critical to the *oxygen cycle,* which keeps the atmosphere from being depleted of its oxygen.) Next the trees are burned, exacerbating the problem by generating huge quantities of additional carbon dioxide. The outcome can only spell disaster—catastrophic climatic changes, glacier meltdowns, massive flooding. It is not a pretty scenario and one that will cut even deeper into the loss of plant and animal life.

Over the past half billion years there have been five mass extinctions. The most recent, occurring at the end of the Cretaceous period about 66 million years ago, was the one that wiped out the dinosaurs. But it was not the worst of these catastrophes. The most devastating loss of life probably took place roughly 245 million years ago, during the great Permian extinction. Between 77 percent and 96 percent of all marine animal species disappeared. Land creatures were similarly decimated.

If the rain forests are completely destroyed, a distinct possibility by the middle of the next century, we just might experience mass extinction number six. As with the blowouts of the distant past, life would probably reassert itself. It is resilient in that way. But gone forever will be the Darwin moth, with its eight-inch tongue perfectly adapted for scooping nectar from a specific variety of orchid. Gone will be the glass frog, with its nearly transparent body and green bones.

Charles William Beebe, the great naturalist-explorer, said it quite eloquently: "When the last individual of a race of living things breathes no more, another Heaven and another Earth must pass before such a one can be seen again."

Chaos Theory:
Predicting the Unpredictable

Have you ever been transfixed by the ever-changing pattern of eddies created by a fast-moving stream? Have you ever watched a plume of smoke rise smoothly from a cigarette, only to break up into a disordered pattern of swirls? When smooth flow in a fluid becomes confused and chaotic, the phenomenon is called *turbulence*—and the beauty of a babbling brook notwithstanding, turbulence is almost always undesirable. It creates drag in pipelines. It destroys lift in airplane wings. Turbulent flow in blood vessels can even interfere with the operation of artificial heart valves. In general, turbulence causes any device that deals with fluids in rapid motion to operate less efficiently.

Unfortunately, turbulent motion of fluids, although studied for centuries, continues to be one of the least understood problems in classical physics. To this day no one really knows why the flow of water in a perfectly smooth pipe, with a perfectly even flow, will turn chaotic if the rate of flow is increased. The most powerful computers are incapable of accurately tracking turbulent flow for more than a few seconds. Werner Heisenberg, the "uncertainty principle" man of quantum theory, declared on his deathbed, "I will have two questions to ask of God: Why relativity and why turbulence? I really think He will have an answer to the first."

God may be stumped, but an emerging new science holds the promise of providing answers. It is the science of *chaos theory*.

Chaos theory is a discipline that attempts to understand why many natural systems are inherently random and unpredictable. The classic example of unpredictability is, of course, the weather. With all their knowledge and computer sophistication, meteorologists still cannot accurately forecast the weather beyond a day or two. Why? Is the problem one of incomplete knowledge of atmospheric conditions? Sir Isaac Newton, when he formulated his classic laws of motion, firmly believed that the universe was deterministic. If one knew the mass and position of matter and the forces acting on it, one could accurately calculate the behavior of that matter for all time. Forces that produce weather by acting on the atmosphere should be no different. If we had all the data, we should be able to produce accurate weather forecasts for all time. Right?

The answer to this immensely important question came in the early 1960s, from a quiet, unassuming meteorologist named Edward Lorenz. Working with a very simple weather-modeling system of only a dozen equations (today's weather models, employing supercomputers, use about a million), he demonstrated that a very minor change in the initial input data—perhaps a temperature reading that differed by a hundredth of a degree—would, over a surprisingly short period of time, produce drastically altered weather. For some reason the equations showed *extreme sensitivity to initial conditions*. It is a phenomenon jokingly referred to as the "butterfly effect" since a butterfly flapping its wings in Tokyo could conceivably cause a hurricane in Florida.

In future investigations Lorenz pared down his twelve equations to a ludicrously simplistic three-equation system, which essentially described convection in a heated fluid. Yet once again, although the equations were deterministic— they described precisely the behavior of the fluid—there was sensitive dependence on initial conditions. Above a certain temperature, slight alterations of heat input produced drastic changes in the eventual motion of the fluid. Perhaps more important, the behavior of the fluid was ever-changing, random, chaotic. One could not predict where a particle within the fluid might be at a later time.

"Even in simple systems which obey Newton's laws of motion, you cannot always predict what is going to happen next." These words, written by Ian Stewart, a chaos expert, pretty well sum up what chaos theory is all about. And the reason for this, in a word, is *instability*—persistent instability. When a system is stable we can write simple linear equations to express its behavior. In other words, the elements in the system behave in proportion to one another. A simple example is the acceleration of a hockey puck along the ice. In the absence of any friction between the puck and the surface it is sliding along, a neat linear equation expresses the relationship between the force applied to the puck and its acceleration:

$$force = mass \times acceleration$$

Doubling or tripling the force merely doubles or triples the acceleration.

Unfortunately, there *is* friction, and that makes the system nonlinear and unstable. As James Gleick explains in his book *Chaos*, "with friction the relationship gets complicated, because the amount of energy changes depending on how fast the puck is already moving. . . . You cannot assign a constant importance to friction because its importance depends on speed. Speed in turn depends on friction."

Trying to figure out a nonlinear system is like trying to play a card game in which the rules are constantly changing. Calculations and predictions of future conditions become less than perfectly accurate. How much less than accurate depends on the degree of nonlinearity, and determining that is a focus of chaos theory. Chaos theory might more accurately be called the *science of nonlinear dynamics*.

When you think about it, you quickly realize that non-linearity is ubiquitous. Most natural systems (such as weather or water flow) are nonlinear. Friction, heat, air resistance, and viscosity are omnipresent, and their introduction into natural systems make them nonlinear. Additionally, any system that is acted upon by more than one force also becomes nonlinear. Figure 1 depicts such a system—a pendulum with an iron bob being acted upon by two magnets. As the bob passes over the magnets in the base, each exerts a force on it. The system is nonlinear. At a point midway between the magnets, each exerts a nearly equal force on the bob. It is here that the nonlinearity of the pendulum's motion is at its greatest. The nonlinearity creates extreme sensitivity to the slightest change in velocity or position of the bob. Given these conditions, the motion of the pendulum will, over a short number of swings, become chaotic, swinging sideways as well as back and forth, totally out of rhythm.

Figure 1
Chaotic Pendulum

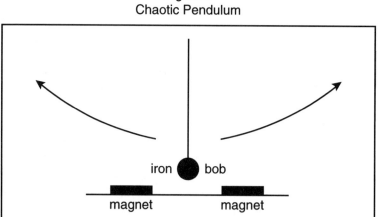

Sensitive dependence on initial conditions is the signature of a chaotic system. But how sensitive is sensitive? Ian Percival, chaos expert and professor of mathematics at Queen Mary and Westfield College in London, explains, using the pendulum and its attracting magnets as an example.

Suppose the sensitivity is so great that error in measuring its position increases by 10 times in one swing between the magnets, which is not at all exceptional. In that case, predicting its position to within a centimeter after one swing entails measuring the [initial] position to within a millimeter. To make the same prediction after four swings, its [initial] position would have to be measured to within the size of a bacterium, and after nine swings, to within less than the size of an atom. The pendulum obeys Newton's deterministic laws, but any attempt to predict its future behavior over long times will be defeated.

If nonlinearity and its offspring, chaos, are indeed everywhere, why did chaos theory not arise until the mid- to late twentieth century? The answer is that scientists in the past tended to ignore the nonlinearities in nature. They wrote up equations that were approximations of the real world, conveniently omitting the nonlinear components. For example, when Galileo rolled balls down inclined planes, establishing the basic laws of gravitational motion, he ignored the minor discrepancies that friction produced. When he dropped heavy and light objects from the Leaning Tower of Pisa to prove that they fall at the same rate of acceleration, he discounted air resistance. The linear equations he formulated were idealized representations of the real world.

For the most part their approach worked. Chaos will not be introduced into a system unless nonlinearity is sufficiently great. And even then it might take quite some time and be insignificantly small. The solar system is a good example. Orbits of the planets around the Sun certainly seem regular. So do orbits of the moons about their respective planets. Yet the solar system is a many-bodied system, each exerting gravitational force on the others. As such it is nonlinear, with the potential to become chaotic. Gleick acknowledges this when he states, "The orbits can be calculated numerically for a while, and with powerful computers they can be tracked for a long while before uncertainties begin to take over. . . . Even today, no one knows for sure that some planetary orbits could not become more and more eccentric until the planets fly off from the system forever." Gerald Sussman and Jack Wisdom of MIT express the view that any calculation of velocity or location of the planets in our solar system would be in gross error after a mere *four million years*.

So, we have nonlinear systems in nature that are inherently unpredictable because predictability demands infinite precision. Does this mean we must resign ourselves to never understanding the world around us? Thankfully the answer is no, and the reason, in a word, is *computers*.

Computers are to chaos theory what the microscope was to biology. The reason is simple. Computers are incredibly powerful and fast—millions of times faster than the human brain. They can put nonlinear equations, those that don't behave proportionally and which we hope reflect the real world, through hundreds of millions of repetitive cycles, called *iterations*, feeding the output of one calculation into the equation as the input for the next calculation. This feedback allows us to watch the system evolve over time. What we see is imperceptible changes becoming greatly amplified.

As an example, here is a nonlinear equation ecologists have come up with to calculate yearly fluctuations in animal populations:

$X' = RX (1-X)$ where X' = the population one year
X = the population the preceding year
R = a constant that represents the number of offspring each adult of that generation produces (or the growth rate of the population)

To make the equation work, X and X' are expressed as fractions between 0 and 1, the real population being 10,000 or 100,000 or 1,000,000 times these fractions.

When a feedback loop is created, in which the population of one year (X) is fed into the equation to get the population for the following year (X'), some very strange things happen, depending on the value of R. Let's set R equal to 1 and see how a population of .4 changes over the years:

First year:	$X' = RX (1-X)$
	$X' = 1 (.4) (.6)$
	$X' = .24$
Second year:	$X' = RX (1-X)$
	$X' = 1 (.24) (.76)$
	$X' = .18$
Third year:	$X' = RX (1-X)$
	$X' = 1 (.18) (.82)$
	$X' = .15$

Each year the population continues to decrease, heading toward zero, or extinction. Extinction, in fact, will occur at any R value of 1 or less, no matter what the initial population. At an R value of 2, the population eventually stabilizes at .5. Higher values of R cause the population to alternate yearly between two different values. This is termed a *bifurcation*. If R is raised still higher, further bifurcations occur, in which the population alternates between four different values, then eight, sixteen, thirty-two, etc. Mysteriously, at an R value of 3.57, the different annual populations become infinite, varying randomly from year to year. Chaos has been reached in the system.

But it doesn't end with chaos. Further R increases produce odd bifurcations and trifurcations, followed by chaos again. What we have is an infinitely deep and complex system—all from a simple nonlinear equation.

The computer is an amazingly powerful and invaluable tool in studying chaos. Not only can its millions of feedback iterations generate long lists of numbers, but it also has graphic capabilities. If each individual solution to an iteration is represented as a point in space, then over many, many iterations of an equation a pattern or picture emerges on the computer screen. The region in space where all possible solutions of the equation fall is called its *phase space*. What evolves over many repetitions is a *phase space portrait* of the system. It is a picture that depicts the long-term behavior of the system the equation represents.

If the equation happens to be linear, the phase space portrait traces a simple curve on the computer screen. But if it is nonlinear, and if chaos is permitted to develop in the system, then an incredibly complex and convoluted tangle of curves is generated, no two curves tracing exactly the same path. This type of phase portrait is known as a *strange attractor* and it is a computer snapshot of a system in chaos. Figure 2 shows the strange attractor Edward Lorenz created with his simple set of weather equations.

Figure 2
Lorenz Strange Attractor

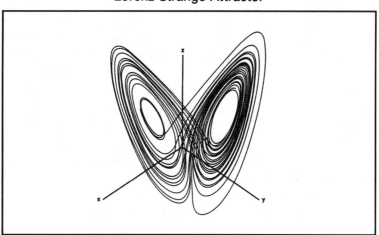

Frank Hoppensteadt of New York University fed the animal population equation through a powerful computer, iterating it hundreds of millions of times at a thousand different values of R. He found a system that was infinitely deep and complex. To quote Gleick on this, "The bifurcations appeared, then chaos—and then, within the chaos, little spikes of order, ephemeral in their instability."

By studying the structure of strange attractors, we are learning a great deal about chaotic systems. Perhaps the most intriguing discovery is that within the randomness of a strange attractor there is order. Benoit Mandelbrot, one of the giants in the investigation of chaos, noticed this when he performed computer iterations on simple nonlinear polynomials. What he got were strange attractors with very complex patterns of curves, which he called *fractals*. But the amazing thing about these fractals was that their complexity was repetitive. Irregularities perceived on a grand scale repeated themselves on ever-decreasing scales within the pattern—a property known as *self-similarity*.

The *Koch snowflakes*, named for its discoverer, Helge

Figure 3
Close-Up of the Koch Snowflake Curve

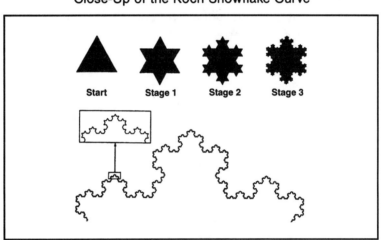

von Koch, is a good example of a fractal. You can easily create it with a pencil and paper by following the sequence of diagrams in Figure 3. First, start with an equilateral triangle. Then add an equilateral triangle hump to the middle third of each side of the triangle. Repeat the process, adding triangular bumps to each new side. What you get is a "frilly snowflake" that can be as intricate and detailed as you choose to make it. And, as the diagram shows, at every level of magnification and inspection, the pattern is self-similar.

Mandelbrot-type fractals have become very popular as an art form. When reproduced in brilliant colors they are, indeed, a spectacular sight. It is, however, in their ability to describe happenings in the real world that the importance of fractals is realized. From the fractal nature of strange attractors we are learning about the infrastructure of chaotic systems and the possible order that lies within them. The turbulence of fluids, the unpredictability of weather, the irregularities of a beating heart, the uncertain nature of an immune response, random fluctuations of electronic systems and chemical reactions, even the ups and downs of the stock market will all be better understood through application of chaos theory. It is an exciting time, and with the creation of ever-more-powerful computers, who knows what secrets we will uncover?

Heavenly Shadows

In 1941 Isaac Asimov wrote a short story called "Nightfall." It launched a legendary writing career and was later voted by the Science Fiction Writers of America as the best science fiction short story ever written. It was about a planet called Lagash, which circled six suns. Celestial mechanics were such that at least one sun always shone in the sky. Darkness never came; night never fell; stars never came out. Then one time it happened . . . the unthinkable. Night did fall, and the stars did come out. And the people of Lagash quietly went mad. But how could darkness descend with a sun ever present in the sky?

Nightfall at Noon

The answer is a *solar eclipse*, which not only happened on Lagash but happens as a matter of course here on Earth as well. Very simply, a solar eclipse occurs when the Moon comes between Earth and the Sun, covering the Sun and casting a Moon shadow on Earth (see Figure 1). (Keep in mind that, unlike the Sun, the Moon does not give off any light of its own. It is a huge mirror, reflecting sunlight that strikes its surface.)

Figure 1
Solar Eclipse

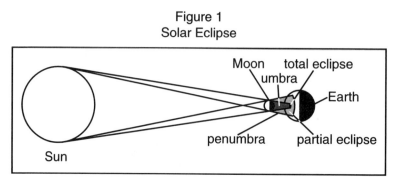

(In all diagrams in this essay, sizes and distances are not drawn to scale.)

You may notice from the figure that the Moon's disk-shaped shadow on Earth has two parts—a *penumbra* and an ***umbra***. The penumbra is a partial shadow; some light from the Sun's surface passes into the penumbra. The umbra, on the other hand, is a zone of total shadow; no light from the Sun's surface passes into the umbra. Any area on earth that is in the penumbra of the Moon's shadow has a *partial solar eclipse*. To an observer it appears as though a piece of the Sun has been cut off. Any area on Earth in the umbra of the Moon's shadow has a *total solar eclipse*. The Sun's surface is completely covered by the Moon. Only a halo of light, the glow from the Sun's atmosphere, can be seen around the dark disk of the Moon.

Solar eclipses are not rare events—not nearly as rare as on Lagash. In fact they occur from two to five times a year, with as many as three of them total *somewhere* on Earth. (Keep in mind that only a small portion of Earth's surface experiences a total solar eclipse, even though the event is classified as one. Most of Earth experiences a partial eclipse or no eclipse.) Yet logic would dictate that they *should* occur every time the Moon comes between Earth and the Sun in its orbital trek around Earth, or about once a month.

To understand the logic, start with a total solar eclipse. The Moon is directly between the Sun and Earth, as viewed from overhead. (When the Moon is aligned with the Sun and Earth in this manner, it is said to be *new*. Eclipses can occur only during a new Moon.) The Moon revolves around Earth in a counterclockwise direction. In $29\frac{1}{2}$ days it will once again be between the Sun and Earth and again $29\frac{1}{2}$ days after that. If there is a solar eclipse every time the Moon comes between the Sun and Earth, why isn't there a solar eclipse every $29\frac{1}{2}$ days? In fact, why isn't there a *total* solar eclipse every $29\frac{1}{2}$ days?

The Proper Alignment

The answer lies in the fact that the Moon is not *really* between the Sun and Earth every $29\frac{1}{2}$ days, or at every new moon. Although it is aligned with the Sun and Earth as seen from overhead, it is usually not aligned with them as seen from the side. It may be above or below. In these instances the Moon does not cover the Sun, and there is no solar eclipse. Sometimes the alignment of the three bodies is partial, leading to a partial solar eclipse.

The Moon does not align with the Sun and Earth at every new moon because the orbit of the moon around Earth is tilted with respect to the orbit of Earth around the Sun. (The tilt is about five degrees.) Most of the time the Moon is above or below the plane of Earth's orbit (see Figure 2). Only when the Moon crosses the plane of Earth's

Figure 2
Bodies Must Align

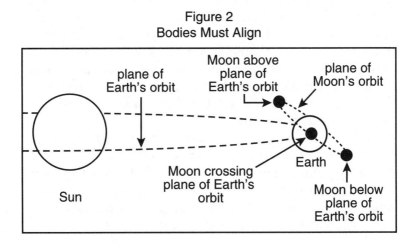

orbit can there be a solar eclipse. And then it must happen at the proper time of the month, when the Moon is new. This combination of requirements—of both space and time—causes solar eclipses to occur much more rarely than every month.

In fact, *total* solar eclipses are even rarer than celestial mechanics would have us believe. The Moon can be aligned perfectly with the Sun and Earth and still not cover the Sun completely. Where we would expect a total solar eclipse only a partial eclipse appears.

A Matter of Size

Size is the determining factor here. For the Moon to cover the Sun completely, it must be at least as large as the Sun is in the sky. By a remarkable coincidence, the two bodies are about the same size when viewed in the sky.

Of course, being the same size in the sky does not mean they are the same size. The Sun has a diameter nearly four hundred times larger than the Moon. But it is also nearly four hundred times farther away from Earth. This makes them *appear* to be the same size, and in the case of eclipses (and, alas, in so many others) appearance is all that counts. Apparent size is measured in the number of degrees of sky an object takes up, referred to as *angular measure*. Both the Sun and Moon have diameters that take up about half a degree. That's roughly the size of a children's aspirin held at arm's length. Your thumb has a thickness of about two degrees, and your fist between eight degrees and ten degrees when held at arm's length. About four moons can fit across the thickness of your thumb. The Sun and Moon both have the same angular measure.

Almost. Sometimes the Moon is a bit farther from us or the Sun a bit closer to us than at other times. (Believe it or not, the Sun is about three million miles *closer* to us in winter than in summer.) On these occasions the Moon may appear a little smaller than the Sun and may not be able to cover it completely, even during a perfect alignment of

Earth, Moon, and Sun. A thin ring of Sun will remain around the dark disk of the Moon. A solar eclipse of this type is called an *annular eclipse*, from the Greek *annulus*, which means "ring." Figure 3 illustrates an annular eclipse. Since no part of Earth's surface is in the dark, or umbral, shadow of the Moon, an annular eclipse is always partial. A spectacular annular eclipse occurred in the United States on May 10, 1994. It was visible in annularity from Texas to Maine, and it displayed its fiery ring for more than six minutes.

Figure 3
Annular Eclipse: Ring of Fire

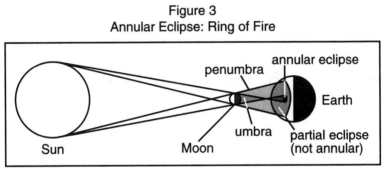

The Total Experience

Of all solar eclipse types the total is unquestionably the most breathtaking. Close your eyes, use your imagination, and together let us experience the grandeur of a total solar eclipse.

It all begins when the Moon makes *first contact* with the Sun, usually along your right-hand side. (First contact can be nearly at the top or bottom of the Sun, however.) As the eclipse progresses, the Moon continues to encroach on the Sun. A noticeable darkening of the day occurs after about half of the Sun has been covered, though it is still clearly daytime. This darkening deepens gradually as the Sun becomes an increasingly thinner crescent. (A *waning* Sun?!) At about 90 percent cover the darkening becomes impressive. It is no longer "just a cloudy day." Something significant is happening. Night is falling. Bats may come out and flowers may close.

With less than a minute to go before total "Sun block," or totality, the Sun shrinks to a mere sliver, then a rim of light against the advancing edge of the Moon. This rim may break up into points of light as the last rays of sunlight shine through the valleys and craters that pock the moon's surface. This necklace of light is called *Baily's beads*, after a nineteenth-century astronomer who described the phenomenon. Just before totality, with perhaps five seconds to go, the necklace may reduce to one single gem of sunlight shining through a particularly deep valley set on a halo of light. You are witnessing the *diamond ring*.

Then totality occurs—also known as *second contact*. It is truly nighttime. On a clear day stars will come out. The outer atmosphere, or *corona*, of the Sun, which has a soft, ghostlike glow of its own that is normally washed out by the brilliance of the Sun's fiery disk, suddenly turns on as if a switch has been flipped. This glow of opalescent light is one of the true spectacles of a total solar eclipse. Also, from around the rim of the Moon solar activity can be seen: red tongues of hot gas flaring up from the Sun's surface and lower atmosphere. It doesn't get any better than this!

Before long—alas!—totality ends. The Sun begins to show its shining face again—usually along your right-hand side (though, once again, it can be toward the top or bottom of the Sun). Astronomers refer to this as *third contact*. Everything that happened before totality now happens in reverse—the shutting off of the coronal glow, the diamond ring, Baily's beads. The stars go back into hiding. Daylight returns. The eclipse ends with what is called *last contact*.

Totality can last up to seven minutes and forty seconds. On average it lasts about $2\frac{1}{2}$ minutes. The entire eclipse, from first to last contact, may last up to about three hours.

Frequency of Occurrence

As already mentioned, solar eclipses are not rare events. However, they occur over a relatively small portion of Earth's surface—especially total solar eclipses. As Earth rotates, the umbral shadow travels across its surface along a *path of totality*. The width of this path is never more than 167 miles (269 km) *at the equator* (at more northerly or southerly latitudes, the shadow strikes Earth obliquely and spreads out somewhat) and usually much less than that. More than 95 percent of Earth's surface falls outside the path of totality and does not experience a total solar eclipse.

What all this means is that although total solar eclipses are not rare, they are rare *at any one place* on Earth. For example, the most recent total solar eclipse seen in *any* part of the continental United States occurred on February 26, 1979. One will not happen again in the continental United States until 2017—thirty-eight years later. It is even rarer for a total solar eclipse to recur in *exactly* the same place, not just the same country or continent. Much rarer—about every 350 years!

A particularly good total solar eclipse—in terms of duration and totality—happened on July 11, 1991. Totality occurred in parts of Hawaii, Mexico, and Central America and was as long as six minutes and fifty-eight seconds— close to the maximum. The last eclipse that equaled or exceeded it was in 1955, and the next one will be in 2132— 141 years later! Table 1 lists solar eclipses for the ten-year period 1991 to 2000.

In addition to partial, annular, and total solar eclipses, there is a fourth type: annular/total. (None occur between 1991 and 2000. However, over the next several decades, one does occur in 2005 and another in 2013.) In such an eclipse the Moon totally covers the Sun only in the center of the eclipse path. Outside the center the curvature of Earth's surface causes it to be a bit farther from the Moon and Sun, which causes the Moon to appear smaller than the Sun— creating an annular eclipse. (In a simple total solar eclipse the curvature of Earth does *not* cause the Moon to become smaller than the Sun outside the center, and the eclipse is total without any annular section. Of course *all* eclipses are partial in the outer, penumbral region of shadowing, as shown in Figures 1 and 3.) Of the different types of solar eclipses, partials are the most common and annular/totals are by far the rarest.

Table 1
A Decade of Solar Eclipses

Date	Location	Type
Jan. 15–16, 1991	Australia, New Zealand, S. Pacific	Annular
July 11, 1991	Hawaii, Pacific, Mexico, Brazil	Total
Jan. 4–5, 1992	C. Pacific, S.W. United States	Annular
June 30, 1992	Uruguay, S. Atlantic	Total
Dec. 24, 1992	Arctic	Partial
May 21, 1993	Arctic	Partial
Nov. 13, 1993	Antarctic	Partial
May 10, 1994	Pacific, Mexico, United States, Canada	Annular
Nov. 3, 1994	Peru through Brazil, S. Atlantic	Total
April 29, 1995	S. Pacific, Peru, S. Atlantic	Annular
Oct. 24, 1995	Iran, India, E. Indies, Pacific	Total
April 17, 1996	Antarctic	Partial
Oct. 12, 1996	Arctic	Partial
March 9, 1997	Mongolia through Siberia, Arctic	Total
Sept. 2, 1997	Antarctic	Partial
Feb. 26, 1998	Pacific, Panama through Guadeloupe	Total
Aug. 22, 1998	Indian Ocean, E. Indies, Pacific	Annular
Feb. 16, 1999	Indian Ocean, Australia, Pacific	Annular
Aug. 11, 1999	N. Atlantic, S.W. England through India	Total
2000	None	—

Blinded by the Light

Caution! Never view a solar eclipse directly without proper eye protection—an approved solar filter. (Smoked glass, x-ray film, or sunglasses *will not do*!) Viewing a partially eclipsed Sun (including annular) for as little as thirty seconds can cause permanent eye damage, which may range from a permanent impression of the Sun superimposed on everything you see to total blindness. (During a solar eclipse in 1970, 140 cases of eye injury were reported in the United States.) If you use binoculars or a telescope, eye damage can occur in *less than a second*. The culprit is infrared radiation, which burns the retina of the eye. And there is no warning; blindness comes painlessly, because the retina has no receptors for pain sensation. The only safe way to view the Sun directly during a solar eclipse is at totality.

The Sun produces infrared all the time, not only during an eclipse, so you might wonder why viewing an uneclipsed Sun does not also cause retinal damage. It would if you could view it long enough, but its brilliance precludes uninterrupted viewing for more than a small fraction of a second. When it's cloudy, you need not worry: clouds block out infrared rays, preventing eye damage.

In the Shadow of Earth

The Moon can hide as well as the Sun, an event known as a *lunar eclipse* (see Figure 4). Solar and lunar eclipses are both the result of shadowing—a solar eclipse is the Moon's shadow on Earth, and a lunar eclipse is Earth's shadow on the Moon—but that is really the extent of their similarity. During a solar eclipse an observer cannot see the Sun because it is physically blocked. During a lunar eclipse the Moon is not physically blocked; it is eclipsed because no light is reflecting off it. (Remember, the Moon is a gigantic mirror.)

Like solar eclipses, lunar eclipses may be partial or

Figure 4
Lunar Eclipse

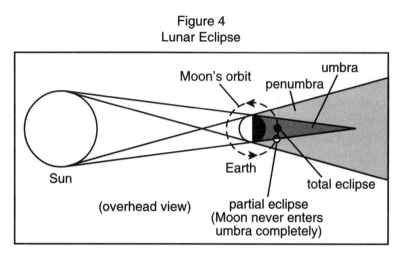

total. But the penumbra of the shadow does not play a role here as it does in solar eclipses. (Technically, *penumbral eclipses*—when the Moon enters only the penumbra of Earth's shadow—do exist, but they are difficult to notice from Earth and not worth going into.) If the Moon is entirely in the umbra of Earth's shadow, the eclipse is total. If it is only partly in Earth's umbra, the eclipse is partial. Unlike a total solar eclipse, a total lunar eclipse is not a local event. Totality occurs at just about all places on the night-time side of Earth—or half the planet.

As Figure 4 shows, a total lunar eclipse occurs only when Earth is directly between the Sun and the Moon—which is the *full moon* phase. It would seem that there should be a total lunar eclipse every time the Moon revolves around Earth into this full moon position—every 29½ days. But it does not happen nearly this often and for the same reason that a solar eclipse does not occur every 29½ days. The Moon and Earth have different orbital planes and do not truly align with the Sun at every full moon.

Lunar eclipses *seem* to be more common than solar eclipses because they are experienced over a much wider area. Any one place on Earth will experience a lunar eclipse much more often than a solar eclipse. Still, the fact is that lunar eclipses occur a bit *less* often than solar eclipses.

Better than two happen per year, on average, and more than half are total. Table 2 lists lunar eclipses for the ten-year period 1991 to 2000.

Table 2
A Decade of Lunar Eclipses

Date	Type (% of Moon darkened)
Dec. 21, 1991	Partial (9%)
June 15, 1992	Partial (69%)
Dec. 10, 1992	Total
June 4, 1993	Total
Nov. 29, 1993	Total
May 25, 1994	Partial (28%)
April 15, 1995	Partial (12%)
April 4, 1996	Total
Sept. 27, 1996	Total
March 24, 1997	Partial (93%)
Sept. 16, 1997	Total
1998	None
July 28, 1999	Partial (42%)
Jan. 21, 2000	Total
July 16, 2000	Total

Lunar eclipses pass through the same general stages, or contacts, that solar eclipses do, but the pace is much more leisurely. Because the diameter of the umbra is nearly three times the diameter of the Moon, *totality can last up to two hours*! The whole event can take four hours. Of course, there are no Baily's beads or diamond ring or coronal glow, and the stars are already out. But a lunar eclipse does have a unique and subtle beauty of its own. Contrary to popular belief, the Moon is not completely dark and invisible during totality. Earth's atmosphere refracts some sunlight—especially wavelengths of red and orange—into Earth's umbra, where the light is reflected off the Moon's surface, giving it a somber, coppery hue.

And there is never any health hazard in viewing a lunar eclipse.

Solar and lunar eclipses are, indeed, two of the heaven's more spectacular displays—examples of nature showing off. But they can be informative as well as lovely. In 450 B.C., a Greek observer, Anaxagoras of Clazomenae, reasoned that because, during a lunar eclipse, Earth's shadow on the Moon was curved in a circular fashion, Earth must be spherical. That was 1,942 years before Columbus! Solar eclipses have been used to measure the light and heat emitted by the Sun's corona. In 1919 one was even used to help prove Einstein's general theory of relativity. (See "Generally Speaking.")

Space and Time Travel

Tooling along in an automobile at a speed of 60 mph (97 km/h)—I hope we're not breaking the speed limit—it would take 3.4 seconds to travel across a football field, 50 hours to travel across the continental United States, and a bit more than 17 days to travel around the world. How long would it take to go to the Moon . . . to Mars . . . to the nearest star . . . to the most distant galaxy? Why, if we have managed to visit all of the planets in our solar system (except for Pluto, which is scheduled for rendezvous early next century), are interstellar and intergalactic travel yet distant dreams?

Is it possible for us to go back in time and see dinosaurs roaming Earth?

Is anything that we see truly happening *now*?

The answers may astound you.

Traveling to Other Planets

Let's get back into our automobile and begin a journey on a cosmic highway that connects Earth to every planet, star, and galaxy in the universe. Kicking the accelerator back up to sixty miles per hour, it would take us about half a year to reach the Moon, ninety-three years to get to Mars at its closest approach to Earth (which happens every twenty-six months), and more than six thousand eight hundred years to reach Pluto, the outermost bastion in the solar system.

Half a year isn't too bad, but who wants to wait ninety-three years to get somewhere? We'd reach Mars only to have our great-grandchildren bury us. As for Pluto at a distant six thousand eight hundred years—our earliest civilizations date back that far. Clearly, sixty miles per hour is not a feasible speed for interplanetary travel. Thankfully we have vehicles that travel much faster.

In 1947 Charles "Chuck" Yeager of the U.S. Air Force broke the sound barrier for the first time, traveling at a speed in excess of 670 mph (1,078 km/h). Six years later he set a new speed record, traveling $2\frac{1}{2}$ times faster. The fastest that any human has traveled since then is 24,791 mph (39,896 km/h), in 1969, aboard the *Apollo 10* command module. Unmanned spacecraft travel even faster. The speediest of all is NASA's German-built *Helios* solar probe, which orbits the Sun and cruises at 115,200 mph (185,391 km/h), or 32 mi/s (51 km/s). (As it rounds the Sun at closest approach, *Helios* can reach speeds of 44 mi/s, but that speed is not constant.) This speed is best suited to our purpose, especially as we travel to increasingly distant places. So let's park the car and hop onto *Helios*. (We can't hop *into* it—it's an unmanned craft.) Traveling at 32 mi/s, it would take about 13 minutes to circle the world, 2 hours to reach the Moon, under 18 days to get to Mars, and $3\frac{1}{2}$ years rather than 6,800 to reach Pluto. Better—much better.

Traveling to Other Suns

Not good enough to get us to the stars, though. The nearest one to us (not including our Sun, of course) is Proxima Centauri. It is twenty-five trillion miles away, or *seven thousand times* farther from us than Pluto is. It would take about twenty-five thousand years for us to reach P. Centauri, traveling aboard *Helios*. And most stars, even within our own galaxy, are hundreds or thousands of times farther away than P. Centauri. Other galaxies are much farther still. Our own galaxy, the Milky Way, contains several hundred billion stars, arranged in a giant spiral. All the stars we see in the sky are part of that spiral. Other galaxies are so distant that we cannot resolve individual stars within them without large telescopes. (Only five of the trillion or so galaxies that pepper the universe are even visible with the naked eye, appearing as faint smudges in the sky.)

Because of the great distances between stars and galaxies, the mile quickly becomes an impractical unit of measure. Instead astronomers use the *light-year*, or *ly*. It is the distance light travels in one year in a vacuum. That distance is roughly 5.9 trillion miles (9.5 trillion km). P. Centauri is about 4.2 ly from Earth. It takes light 4.2 years to travel from Earth to P. Centauri. As already mentioned, it would take *Helios*, which cruises at about *one six-thousandth* the speed of light, 25,000 years. To travel across the spiral of our galaxy takes light about 100,000 years and would take *Helios* about 600 million years. The distances are truly staggering. Can we ever hope to reach even the closest stars in our own galaxy without dying of old age along the way?

Speeding Things Up

Clearly, for interstellar travel even 32 mi/s is not fast enough. We need to reach near-light speed. Nothing can travel faster than light in a vacuum such as space—186,300 mi/s (299,800 km/s). The problem lies in being able to provide enough push, or thrust, over a continuous period of time. The push causes *acceleration*, and for interstellar travel a spaceship would need an acceleration at or near *one gravity*, or *1 g*—the rate at which a freely falling object accelerates toward Earth (disregarding air resistance) due to the gravitational pull between Earth and the object. This rate is 32 feet (9.8 meters) per second per second, or 32 ft/s² (9.8 m/s²). In other words, if a rock is dropped from the top of a tall building, it will have a speed of 32 ft/s after one second, 64 ft/s after two seconds, 96 ft/s after three seconds, and so on, increasing its speed by 32 ft/s every second. If we could get a spaceship accelerating at this rate, it would reach the speed of light after about one year!

But we are not yet able to do that—or even come close. The chemical propulsion rockets that we use today cannot provide the needed thrust over the long haul. The combustion that takes place in the rocket engine is nonnuclear, which burns fuel very inefficiently. It does not pack enough punch. The rocket couldn't even hold enough fuel in its tanks.

Nuclear propulsion systems would be better, and they are in the works. Fission reactors—which release energy the same way atomic bombs and nuclear power plants do—are feasible with present technology. They would be able to accelerate to better than 1 percent the speed of light. The outer planets could be reached in three or four weeks, P. Centauri in four hundred years.

An even better method of nuclear propulsion is nuclear fusion—the kind of energy release that occurs in the Sun and other stars and hydrogen bombs. Unfortunately, present technology is not yet able to release sufficient amounts of energy in a controlled manner by this process. Even more futuristic technologies include methods in which fuel would not have to be carried in the spaceship at all. Take, for instance, a sailing ship—a spacecraft with a sail hundreds of miles across that would tap the energy from the solar wind, a stream of high-energy, high-speed charged particles emanating from the Sun. Theoretically such a ship would be able to attain the 1 g acceleration needed for star travel. There are other, equally fantastic technologies on the drawing boards of dreamers: matter-antimatter engines, for example, or contraptions that could gather and concentrate the sparsely distributed hydrogen atoms of space and use them as a nuclear fuel.

Even if we *could* travel at the speed of light, or darn close to it, we still would be limited in the places we could explore in our lifetime. The most distant stars in our galaxy are 80,000 ly away. The next nearest galaxy to us, the Large Magellanic Cloud, is about 170,000 ly away. The most distant galaxies in the universe are in the vicinity of 15 billion ly away. Even at the speed of light, it would seem that we are bound, by virtue of the vastness of the cosmos, to the nearest few hundred stars in our stellar neighborhood.

Not necessarily!

Traveling Through Time

A puzzling thing happens as a moving object approaches the speed of light. For that object and anything or anyone traveling with it, *time slows down*. It is part of Einstein's special theory of relativity (see "It's All Relative: The Special Theory") and has been borne out experimentally, using atomic clocks. But clocks are not the only things that slow down. Chemical reactions and biological processes slow down as well. The phenomenon is known as *time dilation*.

The relationship between speed of motion and time change is not a simple, linear one. Time does not slow to half its rate at half the speed of light or to one-fourth its rate at one-fourth the speed of light. Speeds must closely approach that of light before time dilation becomes significant (see Table 1). But when it does, wow! At speeds very near that of light, space travelers might age at one-thousandth or one ten-thousandth the rate of their nontraveling counterparts. After a hundred thousand "nontraveler" years have elapsed, our space venturers might have aged a mere ten years. Countless generations would have been born and died; civilizations might have risen and fallen, ice ages come and gone. In a span of ten years! With time dilation, humankind could indeed travel to distant stars and other galaxies—for it is like suspended animation. The stumbling block, once again, is our inability to reach high enough speed.

Not only does time dilation provide a means to explore the universe untethered by our mortality; it is also a kind of time travel. If space travelers can live 10,000 years, are they not traveling into the future, experiencing the future? After

Table 1

Speed of Space Traveler (as a percentage of light speed)	Time-Slowing Factor*
0	1.000
10	1.005
30	1.05
50	1.15
70	1.4
90	2.3
99	7.1
99.9	22.4
99.999	223.6
100	infinite

*Multiply time that elapsed on space traveler's clock by *time-slowing factor* to get time that elapsed on nontraveler's clock.

all, they should be dead 9,920 years "in the past" (assuming a life span of 80 years).

As much as we might like the idea, the phenomenon of time dilation is not truly time travel—becoming *part* of a future time period. Neither is another phenomenon by which we can actually get a firsthand look at the past. We already know that, as fast as light travels, it still takes time. It does not travel instantaneously. For instance, it takes light a little more than one second to travel from the Moon to Earth. Sunlight reaches Earth in about eight minutes. The light from P. Centauri takes 4.2 years to reach us, and the light from a galaxy called M104 takes 40 million years. When we look at the Moon, we see it as it existed a little more than a second ago, the Sun as it radiated eight minutes ago, P. Centauri as it twinkled 4.2 years ago, and M104 (with a telescope powerful enough) as it glowed 40 million years ago. We see everything as it *was*, not *is*. We live in the past. If the Sun suddenly exploded and were gone, we would still see it shining steadily for another eight minutes. We would even be gravitationally bound to it for those eight minutes.

The reverse is equally true. Anyone viewing us from M104 would see Earth as it existed forty million years ago.

How far into the past could we see if our optical telescopes had unlimited capacity? The outer edges of the universe are about fifteen billion light-years away, which means we should be able to see back that far. That was when the universe first began in what has been termed the Big Bang. There is nothing beyond that distance or further in the past than that time. Time, in fact, did not yet exist. Looking back, we would see everything as a blinding white glow, from the light energy that was first being created everywhere in the universe—which, to be technically correct, was about three hundred thousand years *after* the Big Bang. We couldn't actually see past the glow to the *very* beginning. The universe was "opaque" for the first three hundred thousand years. (When astronomers talk about seeing to the "edge of the universe," they do not mean even this far. They mean to the formation of the first stars and galaxies— between one and two billion years after the Big Bang.)

Imagine a planet circling a star in the M104 galaxy. It is a very reflective planet, with a surface made of perfectly smooth and shiny rock. We call it Mirror-X. Imagine also that astronomers of Earth have developed optical telescopes large and powerful enough (not possible yet) to collect and focus light reflected off Mirror-X. Much as you see your own reflection when you look in a mirror, Earth astronomers can see their reflection when they point their telescopes at Mirror-X. Only Mirror-X is forty million light-years away. It takes forty million years for light to travel from Earth to Mirror-X and another forty million years to travel from Mirror-X back to Earth. That's eighty million years.

Astronomers would not see their own reflections; they would see dinosaurs roaming Earth. And I don't mean Jurassic Park!

Again, this phenomenon of seeing things in the past is not time travel. Some theoretical physicists think we can travel back in time—but it is a *very* complicated business, involving rotating black holes, white holes, wormholes, superstrings . . . and possibly more energy than is contained in the whole universe. Certainly not tomorrow's technology but a lot to think about in the meantime.

Alternative Medicine

A patient lay near death. In an effort to cure him, doctors opened his veins and, over a twelve-hour period, let eighty to ninety ounces of blood flow from his body. He was given heavy doses of *calomel*, a powerful and toxic mercury compound, by both mouth and injection. This was followed by *tartar emetic*, a poisonous white salt used to induce perspiration and vomiting. Caustic poultices, which cause blistering, were then applied to various parts of his body. He was forced to inhale vinegar vapors. Finally the patient, struggling to speak, expressed the desire to be allowed to die without interruption.

He did die shortly thereafter, and the medical treatment, which was state of the art, no doubt hastened his demise. The year was 1799, and the patient was George Washington.

Medicine has come a long way since those inglorious days. Many infectious diseases that killed millions—smallpox, diphtheria, polio to name a few—are mere nightmarish memories. Vaccination and antibiotics are, in fact, two of medicine's greatest triumphs. More specifically, they are the triumphs of *orthodox medicine*—medicine that is taught in medical schools, practiced in hospitals, and recognized by health insurance companies. *Alternative medicine* has not fared as well.

Alternative medicine, simply put, comprises all that is not orthodoxy. The more widely accepted alternative therapies include but are not limited to homeopathy, herbalism, nutrition and vitamin medicine, chiropractic, chelation, acupuncture, and acupressure. Lesser lights in alternative medicine's little black bag of tricks are reflexology (foot massage), ozone therapy, ultraviolet blood irradiation (to cleanse the blood), hydrotherapy, yoga, aromatherapy, ayurvedic medicine of India (using mainly herbs and diet to effect cures), and a slew of unclassified techniques. Practitioners of these healing arts have waged a long, uphill battle to prove the efficiency of their specialties.

Part of the problem lies in the counterintuitive nature of alterative medicine. Treatments oftentimes seem totally unrelated to the ills they purport to cure. Why should massaging the sole of your foot relieve a sinus headache (reflexology)? How can sticking slender needles into your skin alleviate, if not cure, irritable bowel syndrome (acupuncture)? Many of the alternative procedures just don't make sense.

Yet tens of thousands of people swear by alternative medicine. It has worked for them. So why not subject some of these questionable procedures to rigorous scientific testing? Easier said than done. To begin with, alternative medicine is in many respects preventive rather than curative. Nutrition and vitamin therapy, for example, aims to keep the body at an optimum level of health so colds, cancers, and the like are unable to gain a foothold. Robert C. Atkins, M.D., a renowned practitioner of alternative medicine, when asked what a patient can be given to relieve his headache, replied, "Instructions on how to make sure he never gets another one."

Perhaps even more important, alternative medicine does not lend itself to strictly controlled experimentation. For example, to gain the approval of the Food and Drug Administration (FDA) a drug must pass what are termed *double-blind* studies. These are experiments involving two identical groups of subjects. One group is given the drug, and the other is given a *placebo*. To eliminate any possible bias, neither subjects nor researchers know which group gets which until the study is concluded; hence the name *double-blind*. It is a valid and time-proved scientific method of investigation.

But how does one subject homeopathy to such study? During an initial visit, a homeopath will spend up to two hours with the patient, eliciting a detailed history. Then he or she will choose a unique combination of remedies tailored to the individual's needs.

Double-blind studies are impossible to perform on most alternative medical procedures (such as acupuncture and chiropractic), just as they are on orthodox procedures such as open-heart surgery. However, the fact that alternative medicine seems to work for a lot of people is good enough for the alternativists. They are what might be termed *empiricists*, trusting anecdotal evidence in the absence of hard, scientific proof.

Despite lack of scientific evidence acceptable to the medical (and health insurance) establishment, in 1992 sixty million Americans shelled out more than $14 billion to alternative therapists, according to a report in the *New England Journal of Medicine*. What did they see in these unorthodox methods that traditional medicine could not provide? Let's take a look at the major therapies of alternative medicine and what they have to offer.

Nutrition-Vitamin Therapy

Regulation of diet to promote good health is something that both mainstream and alternative medicine recognize. We need only look at the low-fat, low-salt diets advocated by the American Heart Association and other health organizations to realize this. There is much, however, in the cookbook of nutrition therapy that orthodox medicine scoffs at. And probably the single greatest disagreement lies in the way the two opposing medical camps view intake of sugar.

Although no one feels that a high-sugar diet is healthful, the nutrition therapists have made sugar avoidance almost a religious practice. According to Dr. Atkins, our unmitigated consumption of sugar constitutes "a national act of health suicide." It can lead, he contends, to heart disease, hypertension, yeast infection (sugar feeds the yeast naturally present in your body, causing these microorganisms to run amok), mental illness, cancer, hyperglycemia (high blood sugar, the extreme of which is diabetes), and a litany of other disorders. As strange as it might seem, eating too much sugar can also lead to hypoglycemia, or low blood sugar. Hypoglycemia and hyperglycemia are really too sides to the same high-sugar-diet coin. Intake of high levels of sugar causes the body to overact in its production of insulin, which then lowers blood sugar to abnormally low levels. And our present intake of sugar is so great that many, if not most, Americans suffer from this insulin-induced hypoglycemia. So say the more nutrition-conscious doctors.

Sugar also robs the body of essential vitamins and minerals by tying them up in reactions necessary to metabolize the excess sugar. For this reason they provide what is often termed *empty calories*. Certain of these reactions convert the sugar into triglycerides, fatty compounds implicated in heart disease. Additionally, the depletion of vitamins and minerals can compromise the immune system, possibly increasing the cancer risk. The bottom line: stay away from sugar.

Another major divide in the thinking of orthodox and alternative medicine is the perceived value of dietary vitamins, especially in megadoses. Vitamin therapy is the main thrust of a larger discipline called *orthomolecular medicine*. The term *orthomolecular therapy* was coined by the Nobel Prize–winning chemist Linus Pauling to describe a protocol that attempts to provide the cells and tissues of the body with all the chemical elements they need to maintain optimum health. Dr. Pauling himself was the champion of vitamin C. Most of the time orthomolecular medicine translates into some form of megadose vitamin regimen.

Vitamins in tiny amounts are essential to proper functioning of the body. Orthodox medicine has written volumes on the subject. Scurvy, beriberi, pellagra, and rickets are each the result of a particular vitamin deficiency. But of what value are megadoses? If the American Medical Association (AMA) or FDA establishes sixty milligrams of vitamin C as the U.S. Recommended Daily Allowance (U.S. RDA), what is the point in taking six hundred milligrams . . . or six thousand milligrams?

Once again, very few double-blind studies have been undertaken to confirm the alleged benefits of megadose vitamin therapy. When tests are done, the results are often inconclusive—as was the case with vitamin C and its alleged ability to cure or prevent colds. One study seemed to suggest that vitamin C can hasten recovery from a cold but has little value in preventing one. Often the dosages used in studies are far below those recommended by vitamin therapists. Many traditional doctors feel that when huge doses of vitamins are ingested, much of the excess is merely excreted in the urine.

Lack of double-blind studies notwithstanding, alternative medicine continues to swear by large doses of vitamins A, C, E, and the B complex as preventive therapy and/or treatment for everything from allergies and the common cold to arthritis and cancer. In truth they are natural antihistamines and anticarcinogens. Alternative medical practitioners are, remember, empiricists. If it seems to work and presents no apparent danger, why not give it a shot?

There is much common ground between orthodox medicine and the practice of nutritional therapy. Not so with the science and medicine of *homeopathy*. It is here that empiricism goes a bit far and reaches some rather fantastic conclusions.

Homeopathy

In 1796 the German physician Samuel Hahnemann took a strong dose of cinchona bark, from which the malaria medication quinine is extracted. He experienced headache, thirst, and fever, symptoms typical of malaria. This led him to a belief that a medication's ability to cure a disease arose from its ability to produce symptoms in a healthy person that are similar to those of the disease. He called it the Law of Similars and used this principle to found homeopathy.

Homeopathy is a medical practice that believes "like cures like." Want to feel better? Take a "bit of the hair of the dog that bit you." If you suffer from yeast infection, then take a pinch more yeast and you'll do just fine. Lead poisoning? Mercury poisoning? What you need is a tad more of that particular heavy metal. Tom Harkin, the Democratic senator from Iowa and chairman of the powerful U.S. Senate Appropriations Committee, helped create the Office of Alternative Medicine because he was convinced that a homeopathic dose of pollen had cured him of his allergies.

Does any of this make sense? Yes and no. Certainly an extra dose of lead should not cure lead poisoning. Yet allergists do treat allergies by a process of desensitization in which small doses of the offending substance are inoculated into the patient. And vaccination works by administering a harmless version of the very germ that causes a particular disease.

Clearly homeopathy cries out for some sort of scientific or statistical scrutiny—scrutiny that has been sorely lacking. This will change, however, now that several studies funded by the Office of Alternative Medicine are being conducted.

Perhaps I present homeopathy in too simplistic a light. Most of the time cures are not as easy to prescribe as in the examples cited here, for the causes of disorder are not easily identifiable. Headache, for example, and also arthritis, fever, and asthma have no readily discernible causes. And if the causes are not apparent, neither are the cures. This is where the expertise of the homeopaths comes into play. Their shelves are lined with hundreds of remedies (about two thousand exist), many of them herbal, and they must determine which combinations of substances work best. The goal is to come up with something that mimics in a healthy person symptoms that are present in the patient—Hahnemann's Law of Similars. Opium, for example, which induces lethargy, profuse sweating, and delirium, supposedly will cure a fever accompanied by these same symptoms.

Most fascinating about homeopathy is its belief in extremely small, extremely dilute doses of medication. How small? How dilute? Consider the following homeopathic preparation of belladonna extract.

Belladonna, also known as *deadly nightshade*, is a highly poisonous plant whose extract causes gastrointestinal distress and breathing difficulty. What better substance to use as a homeopathic medication for people suffering from such problems? Starting with an initial dose of pure extract, the homeopathic remedy is prepared by adding one drop of extract to ninety-nine drops of a water/alcohol solution—a hundred-to-one dilution. One drop of this is then added to another ninety-nine drops of water/alcohol. The procedure is repeated *thirty times*, resulting in a solution a thousand trillion times more dilute than one made from dissolving a single molecule of belladonna extract in all the oceans of Earth. Needless to say, this is so dilute as to contain, in nearly all instances, *not a single molecule* of the substance being administered. Small wonder traditional doctors have trouble taking homeopaths seriously.

Yet for all its seeming worthlessness, tens of thousands of people swear by homeopathy. It is the ultimate empiricism. Some enlist the services of one of the four thousand health-care practitioners (doctors, osteopaths) who include homeopathy in their treatments. Countless others treat themselves with over-the-counter and mail-order homeopathic remedies.

Is the effectiveness of homepathic ministrations merely a placebo effect? Does thinking that a drug will work actually make it so? Perhaps. But in an effort to explain the apparent success of a nonexistent substance, homeopaths have developed several alternative theories. The one most popularly accepted is that vigorous shaking—an important feature in the remedy's preparation—aligns water molecules according to the structure of the active ingredient. When this ingredient is diluted out of the solution, the water retains a "memory" of its structure.

Sound reasonable? I'll stick with the placebo effect.

In one preparation—a homeopathic cure for allergy—several plant and seed extracts were dissolved in a solution that was 70 percent alcohol by volume. That's 140 proof! Maybe now we're getting to the real reason for homeopathy's success.

Chelation Therapy

Unlike homeopathy, *chelation therapy* is quickly gaining respect among orthodox physicians. Not a new procedure, it has in fact been considered the most effective treatment for lead poisoning since 1941. Now the alternative medical community wants chelation to be recognized as an equally effective cure for atherosclerosis, or hardening of the arteries, and it presents a very convincing argument.

Chelation is a chemical process by which heavy metal atoms are bonded to substances called *chelating agents*. In the medical practice of chelation therapy the chelating agent is a synthetic amino acid, *EDTA*, which is introduced into the body and effectively binds to these toxic heavy metals. Once bound to EDTA, the metals are removed from body tissues and excreted in the urine. Metals targeted for removal include harmful pollutants such as mercury and lead, as well as calcium—the mineral responsible for hardening the plaque that clogs our coronary arteries and leads to heart disease.

It is the ability to bind and remove calcium from arterial wall plaque that most interests alternative physicians. They claim that, in more than three-quarters of their chelation therapy patients, coronary artery blockage has been reduced, sometimes by as much as 90 percent. This degree of success makes bypass surgery unnecessary. Why replace clogged coronary arteries, at great expense and inconvenience to the patient, when they can simply be unclogged through chelation?

Practitioners of chelation therapy even go a step or two further in praise of their procedure. They claim that it cleans not only *coronary* arteries but *all* of the body's arteries as well. In so doing, chelation reduces the risk of other vascular-related diseases such as stroke, Alzheimer's disease, and diabetes. One study even showed chelation therapy to lower the incidence of cancer deaths, presumably by cleansing the body of environmental pollutants.

So where's the rub? This chelation stuff seems better than my grandma's chicken soup (although I'm sure EDTA doesn't taste nearly as delicious). Well, there is a downside— or at least there used to be. At one time, early in its use, chelation did pose serious health risks to patients. Deaths occurred as a result of the treatment, due mainly to kidney failure. Microscopic tubules, the filtering structures within the kidneys, were damaged by metallic-EDTA complexes. Today the danger is all but eliminated by lower dosages, slower delivery time, and inclusion of vitamins and minerals to repair damage. In fact, some suggest that washing out the entire vascular system, as chelation does, actually improves kidney function, as well as the functioning of other organs such as the liver, pancreas, and brain.

It has been estimated that more than half a million patients of alternative medicine have had chelation therapy. The treatment is not, however, a one-shot deal. A patient normally requires from twenty to forty weekly or biweekly treatments to sufficiently open his or her arteries. The course of therapy therefore can last for nearly a year. At a cost of about a hundred bucks a pop, we're talking anywhere from $2,000 to $5,000 for the entire procedure. Compare that, however, to the $40,000 for bypass surgery and its expensive follow-up drug regimen.

All things considered, chelation therapy seems like a

very viable alternative to open-heart surgery, which, at a 5 percent annual mortality rate, is not without its risks. Many alternative physicians call chelation "one of medicine's best-kept secrets." Well, it might not be a secret for much longer. Recently, the FDA gave the go-ahead for two clinical studies of chelation therapy, one at Walter Reed Army Hospital, in Washington, D.C., and the other at the Letterman Army Institute of Research, in San Francisco. As this book goes to press, these studies still await funding.

Acupuncture

Do you have a headache? What you need are several very slender needles stuck strategically into your earlobe. Is it your back that ails you? No problem. Just travel up the outer ear a bit and pierce away. The notion of using your ear as a pincushion to effect healing seems outrageous. Yet this is precisely what the ancient Oriental art (or is it a science?) of *ear acupuncture* or *auriculotherapy* is all about. In its more traditional forms acupuncture utilizes the entire body for piercing. Chinese needles are thicker and longer than Japanese needles and are inserted more deeply. Once in place, the needles can be stimulated either by manipulation or by weak pulses of electricity. Laser light has even been used in lieu of needles. In some techniques the needle is merely touched to the surface of the skin and does not penetrate.

Frustrated picadors or legitimate healers? As improbable as the treatment appears, acupuncture is a healing art that cannot be dismissed easily. The reason is simple—it works. Certainly its ability to relieve pain or act as an anesthetic has been well documented. Painful surgeries and dental procedures have been performed on patients whose only form of anesthesia was a dozen or so well-placed needles. And the "power of suggestion" theory can be discounted since similar surgical procedures have been performed on animals, also anesthetized solely through acupuncture techniques.

The ancient Chinese, who perfected this art twenty-five

hundred years ago, defined health as a balance, or harmony, between two opposing sets of forces—the female *yin* and male *yang*. A vital life force called *qi* (pronounce chee) is created as a result of the interplay of the yin and yang. Qi flows throughout the body via fourteen major channels called *meridians*. Disease results from an excess or deficiency of qi in some part of the body. The body is in disharmony, and it is the function of the acupuncturist to restore balance—to rechannel the flow of qi along the proper meridians. This is done through insertion of needles at one or more of the body's two thousand acupuncture points.

As one might expect, Western doctors attempt to explain acupuncture's success somewhat more scientifically. One theory suggests that stimulation of nerves at acupuncture points sends fast-traveling impulses to the spinal cord and brain, which then effectively block pain signals from getting through. Another theory brings the release of natural painkillers called *endorphins* and *encephalins* into play. The first evidence that these neurohumors were being released by acupuncture came from a set of intriguing experiments. When two rats were joined through their circulatory systems, applying an acupuncture needle to one rat increased the pain threshold of both. Clearly some sort of chemical was being released into the bloodstream that affected both rats. When naloxone, a substance that blocks the action of endorphins, was administered, acupuncture was ineffective in reducing pain.

Endorphins are the neurochemicals responsible for producing the "runner's high," a sense of well-being experienced by joggers. Their mode of action is similar to that of the opiates. This might explain acupuncture's success in treating drug addiction—a benefit discovered serendipitously when an opium addict was given acupuncture prior to a surgical procedure. Curiously, his addiction abated. Today a number of detoxification programs are using acupuncture to treat a host of chemical dependencies, including alcoholism. Reports indicate that many of these programs are meeting with at least limited success.

At this juncture in our understanding of acupuncture (I'm a poet!), there is a question which begs to be asked. Acupuncture works well enough in pain management of everything from arthritis to migraines to sciatica, but does it merely mask or alleviate these symptoms without addressing the real problem? Pain is often only the symptom of an underlying problem—the body's way of letting you know something is wrong. Arthritis sufferers claim that along with pain relief acupuncture brings about a reduction in swelling and inflammation of affected joints. Dr. Robert Atkins reports great success at his clinic when using acupuncture (he has a professional acupuncturist on staff) to treat irritable bowel syndrome (IBS). He attributes his positive results to the fact that acupuncture targets the nervous system, which, to a large extent, controls intestinal and colon function. Headache, largely a neuromuscular disorder, also responds well to acupuncture.

But the nervous system is involved in control and coordination of so many bodily functions. Every internal organ is supplied richly with nerves. They maintain heartbeat and blood pressure. Muscle activity and hormone release are regulated by nerve impulses. It is therefore impossible to state with any degree of certainty precisely what the healing limits of acupuncture are. This is a discipline that cries out for more serious scientific investigation. In the meantime, we must be judicious in our use of acupuncture for very serious disorders. Although acupuncture might be considered for the pain management of life-threatening illnesses such as angina or cancer, it certainly would be foolhardy to forsake the more traditional treatments of a cardiologist or an oncologist.

Shiatsu, which means, literally, "finger pressure," is an acupuncture-related discipline. It is, in fact, also referred to as *acupressure*. Along with acupuncture, shiatsu shares the belief that stimulation of certain points within the body (often far removed from the area that hurts) can reduce pain and promote healing. For example, pressing the arch of the foot can heal the kidneys. (Who knows what pressing the little toe can do?) It's all about keeping the vital life forces in harmony.

Chiropractic

Greek—*cheir* (hand); *praktikos* (done by) that science and art concerned with the relationship between the spinal column and the nervous system as it affects the restoration and maintenance of health, primarily utilizing the hands to adjust misaligned and malfunctioning vertebrae.

So reads a plaque that hangs on a wall in my chiropractor's office. Here is another one:

Look well to the spine for the course of disease.
Hippocrates

Propaganda? Perhaps. Yet chiropractic care is probably the only alternative medical procedure that can boast reimbursability from most health insurance carriers. Chiropractors must be doing something right. Let's see exactly what they are doing.

On your first visit to a chiropractor, you will most likely receive several x-rays of the spinal column, from skull to coccyx—or head to tail. This is done to ascertain the extent of misalignment of the vertebrae. In all probability a lecture expounding the virtues of chiropractic will follow. It goes something like this:

Emanating from the spine, through small openings in the vertebrae, are nerves that go to every organ and muscle of the body. When vertebrae are lined up improperly—referred to as *subluxations*—they press or pinch various nerves. In addition to causing pain, these compromised nerves do not send and receive impulses properly. If a particular nerve happens to innervate the pancreas or heart, this organ will suffer some form of dysfunction.

To realign subluxated vertebrae and restore good

health, the chiropractor must perform "adjustments" in which the body is twisted or jerked abruptly. A cracking sound can be heard and felt as vertebrae slip back into place. Although it sounds rather unpleasant, it is not painful.

The theory sounds logical. Unfortunately, there is little evidence to support the contentions of some chiropractors that their adjustments can cure serious organic disorders. Diabetics still require insulin along with any chiropractic adjustments they might be receiving. To date chiropractic care is recognized and recommended by the medical establishment only for relief of pain and discomfort related to the neck and back. Chronic backaches and pinched-nerve disorders, such as sciatica, fall into this category.

The medical community can no longer treat practitioners of nonconventional medicine as snake-oil salesmen. In response to such overwhelming popularity, in 1992 the United States government established the Office of Alternative Medicine, a branch of the National Institutes of Health. Its sole purpose is to study alternative healing methods, such as homeopathy, acupuncture, and chiropractic care. These procedures have resisted serious scientific investigation in the past, and results won't come easy. New innovative methods may have to replace double-blind studies. I wish the office luck.

It's All Relative: The Special Theory

In an essay in *Why Nothing Can Travel Faster than Light . . . and Other Explorations in Nature's Curiosity Shop* (our first book), we said that the greatest scientist of all time was Isaac Newton. That still holds. But moving to modern times, another scientist stands out above all others: Albert Einstein. If not the greatest scientist of all time, he was certainly the greatest scientist of the twentieth century. What he did, very simply, was to rebuild the universe. Newton's universe was a mechanical one, with rules that operated in a fixed and absolute manner. It was a complex and wondrous machine, and scientists—or you and I—could observe the workings of this machine, but its operation was unaffected by and independent of their observation. The observer was, in a sense, outside the machine. Einstein disagreed. The observer, he said, and the conditions under which he observed were critical considerations. Time and space and motion were not absolute but changed relative to the observer. The broad concept is known as *relativity*, and it is perhaps Einstein's greatest achievement. (Curiously, Einstein won the Nobel Prize in Physics in 1921 not for his work in relativity but for explaining the photoelectric effect, in which he changed the world's view of the nature of light.)

131

Relatively Speaking

Imagine that you are taking a car trip. The highway you are riding on is straight, level, and perfectly smooth, and the car is moving at a constant speed of sixty miles per hour (ninety-seven kilometers per hour). You close your eyes. Remarkably, you cannot feel your forward motion. You *feel* as if you are standing still. Perhaps you are—although you certainly would have felt yourself slowing down. Then you reopen your eyes, and you see trees and houses and the road flitting past you, and you are reassured that you are indeed in motion. But the whole thing gets you thinking. Couldn't the car *in reality* be standing still and the world under its tires be moving backward at sixty miles per hour? Though Newton may have pondered the possibility, Einstein said, unequivocally, *yes*. In fact he said that there is no "in reality." For an object moving with uniform motion, all we can say is that the car is moving relative to Earth or that Earth is moving relative to the car or that both are moving relative to each other. (The word *uniform* is important. Uniform motion means that the object is not speeding up, slowing down, or turning. If it *is*, its motion should be detectable in an absolute way. The reasons for this will be addressed in "Generally Speaking.") There is no absolute motion, only motion in relation to another object or system of objects.

Welcome to relativity—or, more specifically, *special relativity*, since it deals with the special case of motion at uniform velocity. (In this book *speed* and *velocity* are used interchangeably, though there are technical differences. Velocity has direction; speed does not.) It was proposed by Einstein in 1905 and consisted of two basic premises:

1. There is no absolute way to tell whether an object is at rest or in uniform motion.
2. Speed of light (in a vacuum, which is presumed throughout this essay) is constant, regardless of the motion of the light source relative to the observer.

The significance of premise 2 and its connection with premise 1 will become apparent in the next section.

Constancy of Light Speed

In the late 1800s and early 1900s scientists wondered long and hard about the relativity of motion. Somehow there *must* be a way to measure an object's absolute motion. But for this to be possible, there must be a frame of reference that one could call *absolute rest*. Any motion could then be measured against that standard. But where could absolute rest be found? It was already known at the time that Earth rotated on its axis (at 1,040 mph, or 1,674 km/h, at the equator) and revolved around the Sun (at 65,870 mph or 106,000 km/h). In addition, and at an even faster speed, the Sun revolves around the center of the Milky Way galaxy. Ten times faster still, the Milky Way whips through the cosmos. It would certainly seem that there is no "at rest" frame of reference that could be used to clock the absolute motion of bodies.

But scientists had other notions. They believed that space was filled with an invisible, intangible substance called *ether* (which means "blazing" in Greek) from a term coined by Aristotle to describe the substance of the heavens. It could not be detected readily, but it *had* to exist to provide a medium in which light waves could travel. Ocean waves need water to propagate them. Sound waves need matter of some kind to carry them along. So, likewise, light must have some material that can transmit its waves. Space, or a vacuum, was not truly empty—it was filled with light-wave-carrying ether.

This ether vibrated, or oscillated, in a wavelike motion, carrying the energy of light through it. But the ether did not have any overall motion or change in position. It never moved. It was fixed throughout the universe, mysterious and undetectable. It simply vibrated to the tune of light energy.

Being fixed in this manner, ether provided the much-needed "at rest" frame of reference. But if it could not be seen, felt, heard, smelled, tasted, or detected in any way, how could ether be used in any practical sense to measure anything?

Enter light. Although ether is undetectable, the light that vibrates the ether is not. It is a kind of *ether tag*. The light waves, in a sense, *are* the ether, much as the undulating movements of water at its surface *are* the water waves, or the vibrations of air molecules following an explosion *are* the sound waves. Light consisted of waves of ether. Without ether there is no light, or no propagation of light.

So, although no object could be measured against another object to determine its motion (you couldn't be sure which one is *really* moving), it certainly could be measured against the fixed ether or, more specifically, the light that is vibrating it. Scientists reasoned (and it was beginning to be shown experimentally) that light moves through the ether at a constant speed. Whether a star is moving toward us or away from us, its speed through the ether is the same. Its vibration of the ether is not influenced by its motion. The speed of light, therefore, could be used as a frame of reference, a signpost, by which absolute motion of an object or an observer could be measured.

If, for example, an observer were traveling toward a beam of light (Figure 1a), the velocity of light measured by the observer should be greater than if the observer were at rest. If the observer were moving away from the beam of light (Figure 1b), the velocity of light measured by the observer should be less than if the observer were at rest.

Figure 1a   Figure 1b

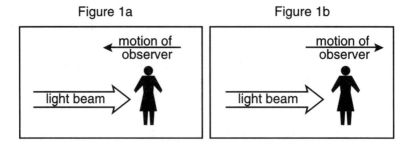

Until the end of the nineteenth century this reasoning could not be put to the test. Instruments were not yet available that were sensitive enough to measure the motion of an object relative to the ether. The change in the velocity of light based on such motion would be infinitesimal. However, in the 1880s, using a newly developed piece of apparatus called an *interferometer*, physicists Albert Michelson and Edward Morley were finally able to test the ether hypothesis. The results were shocking. In all of their experiments Michelson and Morley found no difference in the speed of light. The rotation of Earth, or the motion of *any* object in *any* direction relative to a beam of light, did not change the measured velocity of that beam. The speed of light was constant, not only with respect to motion of the source but also with respect to motion of the observer. It is now recognized as an absolute—one of the very few absolutes in Einstein's universe. Light speed is 186,291 mi/s (299,792 km/s) and is represented by the symbol c.

The constancy of light speed forced scientists to question seriously the ether-filled-space concept and in turn the ability to determine absolute motion. It also seemed a violation of basic logic and common sense. If a pitcher throws a ball toward home plate with speed p, and a batter swings his bat at the ball with speed b, the effective speed of the ball as the two meet (or pass each other, if I'm the batter) should be $p + b$ (neglecting air resistance and gravity's downward tug). It is a simple matter of adding velocities. Why should it be any different if the pitcher "throws" a beam of light instead of the ball? Should the principle not be the same? If the speed of the bat is the same, b, and the speed of light is c, the measured velocity of the beam of light as it meets or passes the bat should be $p + c$—which of course would be greater than the speed of light. Why is the measured velocity not additive?

Einstein reasoned that perhaps the effect of the bat on the speed of the ball is less than we suspect, that it is *slightly less* than b. As the speed of the ball (or any moving object) increases, the effect of the bat continues to decrease. For objects moving at everyday speeds, such as thrown balls, Amtrak (but not during rush hour!), or even our fastest spaceships, this decrease is negligible. The velocities would seem to be additive. But as velocities approach light speed, the decrease becomes significant. At the speed of light the effect of the moving bat on total velocity reduces to zero. Hence, *b* + *c* becomes 0 + *c*, or simply *c*, the speed of light. These assumptions are backed up by rigorous mathematical analysis.

How Long Is It?

In light of Michelson and Morley's findings (no pun intended), and in a desperate attempt to retain the ether-filled-space concept and absolute motion, Irish physicist George FitzGerald proposed that all objects grew shorter in the direction of their absolute motion. He worked it out mathematically so that the extent of shortening exactly accounted for the null results of Michelson and Morley. The concept of length shortening with motion is known as the *FitzGerald contraction* (or the *FitzGerald-Lorentz contraction*, after Dutch physicist Hendrik Lorentz, who expanded the hypothesis). The contraction states that an object traveling at velocity *v* will shorten by the following value:

$$\sqrt{1 - \frac{v^2}{c^2}}$$

Despite FitzGerald's valiant efforts, the ether concept and absolute motion were ultimately doomed. Not so for his notion of length change. Ironically, the shortening of an object with an increase in motion is an important consequence of relativity, which FitzGerald tried to disprove. The shortening occurs in the direction of motion only (*relative* motion, of course) and is significant only at near-light speeds. In fact an object must travel at about seven-eighths the speed of light to shorten to half its rest length. At half the speed of light it would shorten only about one-eighth. Keep in mind that our fastest spaceships reach speeds of about *one five-thousandth* the speed of light. Length shortening at this speed is barely measurable. The shortening of Earth's diameter at the equator due to its motion about the Sun, at one ten-thousandth the speed of light, is a mere 205 feet (62.5 meters)—two-thirds of a football field! Earth has an equatorial diameter of about 140,000 football fields.

The fact that shortening occurs only in the direction of motion has interesting effects. A six-foot-tall man standing on a horizontally moving platform as it approaches light speed would get continually flatter but would remain six feet tall. At the speed of light he would thin to nothing. (It beats Weight Watchers.)

How Massive Is It?

Length is not the only measure that changes with velocity; so does mass. This is another consequence of special relativity. Mass is the amount of matter or substance that a body has. It is often viewed interchangeably with weight, though they are not the same thing. (Weight is a measure of the gravitational pull on a body. On the Moon you would weigh about one-sixth as much as you do on Earth, due to the Moon's weaker gravity. In deep space you would be weightless. Your mass, on the other hand, would not change.)

As velocity increases, the mass of an object increases as well. A golf ball has less mass when it is resting on a tee than when it is flying through the air. You have less mass when you are sitting down watching TV than when you are running to the refrigerator for a snack during a commercial. As with length, though, changes in mass at these *very* low speeds are barely measurable, even with our most sophisticated technologies. At near-light speeds, however, mass increase does become significant. Because mass is increasing with an increase in velocity (the opposite of length change), the mass-change formula is the inverse of the FitzGerald contraction, or

$$\frac{1}{\sqrt{1-\frac{v^2}{c^2}}}$$

For mass to double, velocity must increase to seven-eighths the speed of light. No vehicle or large object can travel nearly that fast, but subatomic particles can. *Particle accelerators* have sped up electrons to velocities greater than 99.9999999 percent that of light. Their mass increased more than forty thousand times. At the speed of light their mass would become infinite.

Why should mass increase along with velocity? Let us view it this way: Energy is imparted to a body, causing it to increase its motion. At speeds well below that of light, this energy is converted almost entirely into motion. As light speed is approached, however, more and more of this energy is converted into mass, with less being turned into motion. *At* the speed of light, all the energy that is added to a body is converted into mass, none into motion. The body cannot speed up any further, and its mass is infinite.

Mass-Energy Conversion

The preceding conceptualization hints strongly at a link between mass and energy. In classical physics matter and energy were separate and distinct entities. Mass took up space; energy did not. Given a gravitational field, mass had weight; energy did not. Special relativity does not make this distinction. It views mass and energy as equivalent, two sides of the same coin. One can be transformed into the other, much as electrical energy can be transformed into heat energy or heat energy into electrical.

Einstein's famous equation, $e = mc^2$ (where e = energy, m = mass, and c = light speed), clearly shows this equivalence. It is, in effect, saying *energy = mass*. The speed of light is added as a conversion factor to indicate how much energy can be gotten from a given amount of mass. It is enormous. All the energy used by mankind in an entire year has a mass equivalent of only a few tons—that of a small elephant.

Conventional or nonnuclear chemical reactions convert *very little* mass into energy. For example, when a gallon of gasoline burns in a car engine, allowing the car to travel perhaps thirty miles (forty-eight kilometers), only *one thirty-thousandth* of one drop of gasoline is converted into energy. The rest remains as mass—the products of the burning process. In general the loss of mass to energy in nonnuclear reactions is too small to measure. In fact scientists were reasonably certain that there was no loss of mass in chemical reactions. Hence they formulated the law of conservation of mass, which stated that mass could neither be created nor be destroyed, merely changed around. Paper could burn, but the ashes and smoke that formed would weigh the same (have the same mass) as the paper did.

Then the nuclear age dawned. It started in 1896 with the discovery of radioactivity and proceeded to nuclear fission (as seen in atomic bombs and nuclear reactors) and nuclear fusion (as seen in hydrogen bombs and the source of energy from stars). In all of these instances the reactions were nuclear. Changes occurred in the nucleus of atoms rather than in the surrounding electron shells. Vast amounts of energy were released, and the loss of mass was easy to measure. Its conversion into energy invariably obeyed Einstein's equation, $e = mc^2$.

With these nuclear-age developments, the law of conservation of mass had to be restated as the law of conservation of mass-energy. Mass *could* be destroyed, or, more correctly, converted into energy, but the total mass *and* energy in the system must remain the same.

The fantastic release of energy as mass that is lost in nuclear reactions helped explain a scientific conundrum: the source of the Sun's enormous energy output. As far back as the mid-1800s geologists realized that Earth and the Sun were at least several hundred million years old. Yet by all calculations based on conventional burning of fuel, the Sun could not be nearly that old. The eminent physicist Lord Kelvin estimated the age of the Sun to be no more than thirty million years.

Today we know that the Sun is 4.6 *billion* years old. What kind of process could account for the Sun's vast energy output yet not cause it to run out of fuel in all that time? Nuclear fusion—with the conversion of mass into energy as described by e = mc². In fact "burning" fuel at its present rate, the Sun will keep going for at least *another* 4.6 billion years before exhausting itself.

An interesting thought: Is the reverse possible? Can energy be converted into mass? Yes, it can. It has been done on a small scale in laboratories in high-energy particle accelerators. (See "What's the Matter with Matter?" for more on these contraptions.) A very small amount of mass has been formed from a very large amount of energy, once again according to e = mc². It also happened about fifteen billion years ago, on a vastly larger scale, when mass formed from an unimaginably dense, expanding ball of energy. It was the creation of the universe.

What Time Is It?

Another, and perhaps the most intriguing, aspect of special relativity is the slowing of time with an increase in velocity. The phenomenon was discussed in "Space and Time Travel" and referred to as *time dilation.* That it exists has been proved empirically by placing very accurate timing devices called *atomic clocks* in jets and rockets and comparing them, after their journey, with identical clocks that had remained Earthbound. The clocks that had "traveled" recorded the passage of less time than the Earthbound clocks. Time had slowed down. The faster the vehicle moved, the more it slowed. As with length change, the rate of slowing is described in FitzGerald's contraction.

As we have already seen by applying this formula, change does not become significant until near-light speeds are reached. At one-tenth the speed of light, a clock would slow by only *.5 percent.* At seven-eighths the speed of light, time would slow down by half. *At* the speed of light, time would stop. Our fastest spaceships (which can travel better than forty miles per second) slow time down by two-millionths of 1 percent—less than one second of slowing in more than one and a half years!

The concept of time slowing would seem to present a contradiction to special relativity. Let us take, for example, what physicists called the *twin paradox* or *clock paradox*. Mary and Betty are twin sisters. On their thirtieth birthday Mary decides to take a trip to Sirius A, the brightest star in the sky. The space cruiser she goes on travels at seven-eighths the speed of light. At this speed time slows down for her by half, relative to Earthbound clocks. The round trip takes about twenty years, Earth time. When Mary returns home, she finds that her twin sister has a few more gray hairs and wrinkles than she does. After all, Mary is only forty years old, while Betty is fifty. Clocks would also show this time difference.

But if motion causes a slowing of time, couldn't this be the long-sought-for frame of reference to determine absolute motion? After all, if Mary is younger than Betty, or her clock shows the passage of less time, then *she* was moving, not Betty. Motion is not relative but absolute.

To confound matters further, according to special relativity it could be argued that Mary's spaceship is stationary while Earth and the rest of space are making the long journey away from the spaceship at seven-eighths the speed of light. In this case time for Earthbound Betty would slow and she would be the younger sister by ten years.

Clearly this is impossible. Each sister cannot be younger than the other. It would certainly seem that if time dilation happens—and it does—special relativity is in big trouble. Philosophers and scientists debated the issue at great length.

In the final analysis, special relativity was preserved—along with Einstein's reputation. You see, neither Mary nor Betty is aware of any time change. Each sees the passage of time normally from her frame of reference. For them to detect any change, they must bring the clocks together and compare them. The only way to do that is to have one of the clocks turn around and come back to the other clock or stop and reverse itself. In either case the clock would be *accelerating*. (According to scientists, acceleration is a change in speed *or* a change in direction.) Uniform motion, on which special relativity is based, implies *straight-line* motion at *constant* speed. Once Mary's spaceship or Betty's Earth turns around to come back home, special relativity no longer applies.

But general relativity does.

Generally Speaking

If special relativity established that Einstein was a great scientist, general relativity marked him as one of the greatest. It was once said (with some exaggeration) that no more than twelve people in the world could understand general relativity. It was subtler and more elegant than special relativity and more sweeping in scope. Special relativity redefined motion; general relativity redefined gravity, space, and time. It restructured the universe.

Special relativity applied to uniform motion only—or motion in a straight line at constant speed. It did not apply to nonuniform or *accelerated* motion—motion in which an object is speeding up, slowing down, or turning. Einstein wondered about this. Why should relativity apply to one kind of motion and not another? He was a simple man who looked for simple answers, and exceptions bothered him. He contemplated the dilemma of nonuniform motion for ten years and in 1915 came up with general relativity.

Gravity and Inertia

The problem with nonuniform motion is that it produces effects that can be felt. If two cars, A and B, are traveling past each other in opposite directions with uniform motion, it is impossible to say which is moving past which. However, if one of the cars is in nonuniform motion, it must be speeding up, slowing down, or turning. If it is speeding up, a passenger inside will be pushed backward; if it is slowing down, the passenger will be pulled forward; if it is turning, he or she will be thrown to the side. These effects of acceleration indicate motion in an *absolute* sense. So much for relativity?

Not exactly. The action of being pushed or pulled or thrown to the side is known as an *inertial effect*. In classical physics *inertia* is the resistance of a body to an applied force. There is no effect of inertia on a body traveling in uniform motion. Barring friction, it will travel in uniform motion indefinitely. A Nolan Ryan fastball thrown at ninety-eight miles per hour in outer space will continue to travel at ninety-eight miles per hour forever. However, if an outside force (such as Frank Thomas's bat) is applied to the ball, speeding it up, slowing it down, or turning it, inertial effects come into play; the ball will resist the force. A passenger's being pushed backward, pulled forward, or thrown to the side is a manifestation of this resistance. It is an inertial effect.

Inertial effects manifest themselves in other ways as well. Have you ever taken a bucket of water and swung it around in a vertical circle? When the bucket is upside down, above your head, the water should spill out, but it doesn't as long as you continue to swing the bucket. The water is pushed outward by the circular (nonuniform) motion of the bucket. This push is called *centrifugal force*, and it keeps the water inside the bucket. (To a physicist there really is no force pushing the water outward. The only real force is exerted by the bucket on the water, pulling it toward the center of the circular path, keeping it in circular motion. However, *to the water*, the centrifugal or outward-directed force is very real and useful to work with for our purposes.) It is a force that acts like gravity, but outward from the center rather than toward the center. It is an inertial effect.

As another example of inertial effects, imagine that you are inside an elevator in deep space, where there is no gravity. The elevator is speeding up in the direction indicated in Figure 1.

Figure 1
Elevator in Accelerated Motion in Deep Space

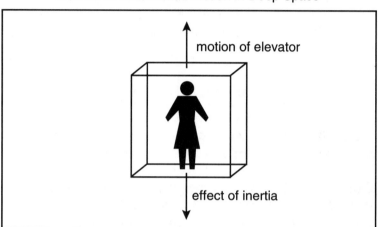

As a result, you are pushed against the floor of the elevator—an inertial effect. Yet it feels as if a gravitational force is holding you down. You jump up and fall back to the floor. You have weight—which is the effect of gravity on a body. If the elevator is accelerating at 32 ft/s² (9.8 m/s²), the inertial effect would be the same as the gravitational force of Earth on the elevator at or near Earth's surface. (Objects near Earth's surface fall with an acceleration of 32 ft/s².)

Einstein reasoned that since the effects of inertia and gravity feel the same, they are the same. *Gravity and inertia are two words for the same thing. They are equivalent.* Therefore, it could be argued that the elevator is not moving upward through the universe, creating an *inertial* effect on the passenger, but that the universe is moving downward past the elevator, creating a *gravitational* effect. (Since the elevator is not moving, the effect can hardly be called inertial.) It is a matter of semantics. Let us take again the situation in which two cars, A and B, are moving past each other. If a passenger in car A experiences inertial effects (such as being pushed backward), but one in car B does not, it simply means that car A is moving *relative to the universe*, and car B is not. It does not establish the *absolute* motion of car A. *All* motion is relative—nonuniform as well as uniform—and can be described by the same set of equations. Inertial effects cannot be used as a frame of reference to determine absolute motion.

Space and Time

General relativity offered a new way of viewing gravity. It also offered a new way of viewing space and time. In classical physics space exists in three dimensions. The simplest way to view those dimensions is to visualize a box. The size of the box can be gotten by measuring its width, height, and depth. Each of these measurements represents one of the three axes, or dimensions, in space. Any place inside the box, or in space, can be described by using these three axes. In such a system there is literally no "space" for a fourth dimension. In mathematical constructs mathematicians *can* create four-, five-, and even greater-dimensional space, but not in the real world. In Einstein's world, however, there *are* four dimensions, and it is a very real world indeed. His fourth dimension is not one of space but of *time*, and his world is *spacetime*. According to Einstein, space and time are inseparable; they are the very fabric of the universe, which has been described as a *spacetime continuum*. (Hermann Minkowski, a Polish mathematician, was actually the first to mathematically describe the concept of time as a fourth dimension. Einstein drew on Minkowski's conclusions to advance his own work.)

It is very difficult to conceptualize *anything* in four dimensions. Calling the fourth dimension *time* instead of another dimension in space doesn't really help. To make visualization easier, imagine the space structure to be two dimensions and time to be the third. The dimensions of our spacetime universe can now be represented once again by a box (see Figure 2). If an object is stationary, it does not move through space but still moves through time—as indicated by arrow A in Figure 2. Movement of the object through space as well, with uniform motion, is indicated by a straight line—arrow B. Movement with nonuniform motion is indicated by a curved line (dashed for easier identification)—arrow C. These representations of movement through spacetime are known as *world lines*.

Figure 2
Two Dimensions in Space, One in Time

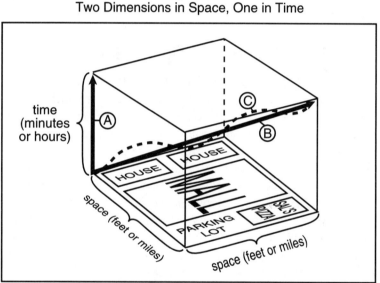

Einstein and his "relatives" were able to explain certain special relativity phenomena, such as length shortening and time slowing, based on the concept of spacetime. Since space and time are not absolutes that exist independently, one can change relative to the other. Martin Gardner, in his illuminating book *The Relativity Explosion*, uses an interesting analogy: A book is held in front of a light, with its shadow projected on a wall. Rotating the book will alter the shape of its shadow. "With the book in one position,"

Gardner says, "the shadow is a fat rectangle. In another position it is a thin rectangle."

The book, of course, does not change its size or shape; only its projection changes. "In a similar way," Gardner continues, "an observer sees a four-dimensional structure . . . say, a spaceship, in different projections depending on his motion relative to the structure. In some cases, the projection shows more of space and less of time; in other cases, the reverse is true." The projection changes as the ship "rotates," or changes its relative motion, in spacetime. This changing projection manifests itself as a relativity phenomenon, such as a shortening of length or a slowing of time.

Curving of Spacetime

In classical physics gravity is a force. It exerts a pull on objects near it. This pull causes a raindrop to fall to the ground and the planets to circle the Sun. Relativists, as we have already seen, do not view gravity in a classical way. According to Einstein, it is not a force at all but an effect. Gravity does not move matter; it curves space—or, rather, spacetime. The more massive an object is, the stronger is its gravity, and the more it curves spacetime. A planet does not orbit the Sun because the gravity of the Sun pulls on the planet, causing it to move around the Sun in an elliptical path, but because spacetime around the Sun curves in such a way that the planet, in traveling along its world line, takes an elliptical path around the Sun. Relativists often use the following analogy, as described in *The Relativity Explosion*, to concretize the concept:

"Imagine a rubber sheet stretched out flat like a trampoline. A grapefruit [the Sun] placed on this sheet will make a depression. A marble [one of the planets] placed near the grapefruit will roll toward it. The grapefruit is not 'pulling' the marble. Rather, it has created a field of such a structure (the depression) that the marble . . . rolls toward the grapefruit," ultimately assuming an elliptical path around it. To quote Einstein: "Space tells matter how to move; matter tells space how to curve."

One additional comment: As gravity curves or warps spacetime, not only is the observed path of an object changed as it is projected onto the dimensions of space, but the observed passage of time is changed as the object is projected onto the dimension of time. *As gravity increases, time slows.*

Is General Relativity for Real?

For years after Einstein advanced his theory of general relativity, physicists doubted its validity. There was abundant evidence for the special theory: the constant speed of light, the shortening and increase in mass of an object as it speeds up, time dilation, the conversion of mass into energy (or vice versa) according to $e = mc^2$—all confirmed through indisputable experimentation. But general relativity remained an experimentalist's nightmare. To quote Misner, Thorne, and Wheeler from their book *Gravitation*: "No theory was thought more beautiful, and none was more difficult to test."

An important test came in 1919, during a total eclipse of the Sun. General relativity proposed that a powerful gravitational field, in warping spacetime, would bend a beam of light that passed very close to the field. A total solar eclipse offered the ideal proving ground. The Sun generated a fairly powerful gravitational field, and starlight, passing close to the Sun on its way to Earth, could be studied for any deflection. A total eclipse was required because otherwise the blinding glare of the Sun would completely wash out the feeble rays of starlight. There was much pre-eclipse publicity, and the reputation of Einstein as well as the fate of general relativity hung in the balance. The results "generally" agreed with Einstein's predictions. The slight deflection of starlight was in the ballpark of his calculations.

Another test of Einstein's general theory came by way

of accurately describing Mercury's orbit around the Sun. It involved a small problem that existed with Mercury's orbit: it did not follow the exact path predicted by classical physics. When the gravitational effects of the Sun and other planets on Mercury were plugged into the equations of classical mechanics, a small shifting in the orientation of Mercury's orbit should occur—about one-seventh of one degree per century. Mercury's orbit, however, changed a bit more rapidly. Astronomers conjectured that there might be an undiscovered planet circulating inside the orbit of Mercury whose gravity would account for this discrepancy. They even gave the planet a name: Vulcan. (That was *before* "Star Trek.") Throughout the latter part of the 1800s there was a feverish search of the heavens for Vulcan, but it could not be found, and never was. Mercury's orbital anomaly remained a mystery until Einstein came on the scene. His general theory solved the mystery. The additional gravity did not come from a tenth planet inside Mercury's orbit but from the Sun itself. To quote Isaac Asimov from his essay "The Planet That Wasn't":

> The Sun's enormous gravitational field represents a large quantity of energy and this is equivalent to a certain, much smaller quantity of mass. Remember, energy and mass are interconvertible, according to $e = mc^2$. Since all mass gives rise to a gravitational field, the Sun's gravitational field, when viewed as mass, must give rise to a much smaller gravitational field of its own.

This smaller *additional* gravitational field curved spacetime just the right *additional* amount, accounting perfectly for the difference in Mercury's observed and Newtonian-calculated orbits.

As mentioned earlier, general relativity also proposes that gravity slows down time. The intenser the field of gravity, the greater the slowing. Using very accurate atomic clocks, scientists have measured this time slowing. In fact, at the top of Mt. Everest (weaker gravity), a clock ticks three-billionths of a second faster every hour than a clock at sea level (stronger gravity). In outer space (no gravity) it ticks even faster.

The slowing of time also has the effect of causing a shift in the spectrum of light toward the red end (called a *redshift*). This shift, to be sure, is very slight for stars such as the Sun, which have a powerful gravity compared to the less massive planets, but not nearly as powerful as gravity can get. The Sun's redshift was confirmed with good accuracy in 1962.

General relativity has been proved far beyond any doubt or question and has become an integral part of scientific thought.

Black Holes, Gravity Waves, and Gravitational Lenses

General relativity predicted or accounted for certain other phenomena that have been borne out in the real world. One of the strangest of these predictions is black holes. As we already know, general relativity postulates that gravity warps spacetime. This warping causes matter and even energy, such as light, to follow a curved path toward the center of the gravitational field. (In Newtonian physics gravity does not influence energy.) This curving may be significant on matter, but is generally very, very slight on energy. A beam of light passing close to a planet or the Sun, or even a star several times more massive than the Sun, would be curved, or deflected, very little. However, a gravitational field of sufficient intensity could be created (by the compacting of matter), with a spacetime warping so great that light would not only curve *toward* the center of the gravitational field but *into* it so that it could not escape (hence the term *black* hole). The light would not be deflected but would be sucked in—a kind of bottomless funnel in which all matter and energy would spiral down to ... who knows where? Matter that enters the black hole would accelerate to light speed and squeeze to infinite density. Time, for all concerned, would slow to a halt. Definitely weird stuff!

Does such weird stuff *really* exist? Astronomers are certain that it does. Confirmation of black holes came on Wednesday, May 25, 1994, when scientists, using the Hubble Space Telescope, discovered an intense gravitational field in the center of a giant galaxy fifty million light-years away, called M87. We know it is a black hole because it is holding together a whirlpool of hot gases that only the gravity of a black hole can do. The evidence, though conclusive, is necessarily indirect: no light or other radiation can ever escape a black hole to send a message of its presence back to us.

There is other indirect evidence for the existence of black holes. Before entering the point of no return (called the *event horizon* or *Schwarzschild radius*), it is believed that matter would emit energy in the form of intense x-ray radiation. There are several powerful x-ray emitters "out there" that astronomers suspect are black holes. They believe that the energetic centers of most large galaxies may be black holes. (They're not sure about our own, which is on the small and quiet side.)

More direct evidence of black holes may be possible with the detection of *gravity waves*—a remarkable prediction of Einstein's general theory. Gravity waves are ripples in the fabric of spacetime, caused by gravitational energy. They travel outward in all directions from their source, which could be anything that disturbs or warps spacetime in a significant way. Black holes warp spacetime in a *very* significant way—and should send out gravity waves with measurable and distinguishing characteristics.

Have gravity waves actually been detected? No. Since the mid-1960s elaborate gravity-wave experiments have been designed and detectors built, but no *confirmed* results have been gotten so far.

Far less theoretical than gravity waves are *gravitational lenses*. They do exist. They were discovered in 1979 and come right out of general relativity theory. Very simply, they are warps in spacetime caused by a vast concentration of mass. Galaxies create gravitational lenses. Matter that cannot be "seen"—called *dark matter*—because it does not generate or reflect any light or other radiation may create a gravitational lens. Such lenses can split light from background stars, galaxies, or quasars into two, three, or even four separate images. They can distort the shape of galaxies into weird arcs or rings of light.

Gravitational lenses have many potential uses. It is believed that as much as 99 percent of the matter of the universe may be dark. But the gravitational effects of dark matter—when it acts as a gravitational lens, for instance—enable scientists to establish its presence and calculate its mass. Knowing how much matter there is can help us determine the fate of the universe. Will it expand forever, or will it stop expanding and contract into a superdense point?

Much of general relativity supports Newton and his laws of mechanics and gravitation, yet much of it extends beyond and sees the universe in a wholly new and different way. It is small wonder that Einstein's accomplishments prompted him to write toward the end of his life: "Newton, forgive me." Yet no forgiveness was necessary. Newton said of his own remarkable achievements: "If I have seen farther . . . it is by standing on the shoulders of giants." So Einstein stood on the shoulders of science's greatest giant.

Forensic Medicine: A Short Course

Part 1: The Medical Examiner and the Autopsy

The Medical Examiner

Homicide is the leading cause of death for Americans between the ages of fifteen and twenty-five; it is the fourth leading cause of death for all Americans under sixty-five. Two million Americans alive today will eventually be murdered. A woman born in the United States today has a 12 percent chance of being forcibly raped sometime in her life. About two million deaths occur annually in the United States, 8 percent of which are unnatural—suicide, homicide, or accident. In people under thirty-five years of age the figure rises dramatically—to more than half.

Such are the statistics. They pose a huge—and growing—challenge for the science of forensic medicine, the science charged with the responsibility of investigating unnatural deaths.

Webster's Unabridged Dictionary defines forensic medicine as "the science concerned with the relations between medicine and the law; medical jurisprudence." At the helm of this discipline is the *medical examiner* or *ME*. Today it is a highly skilled and respected position held by medical doctors who are pathologists well trained in studying unnatural deaths. All obvious homicides are investigated by an

155

ME, although most accidental deaths and suicide are not. It was a medical examiner who first drew attention to the battering of children. But things were not always this way. Let's go back to England roughly eight hundred years ago, to the inauspicious beginnings of forensic medicine.

First mention of the British office of *coroner* (forerunner to the ME and often used synonymously) is found in ordinances dating back to 1194. This should not be surprising given the long and storied history of jurisprudence in England. Somewhat disconcerting, however, is the fact that the primary concern of the coroner in medieval times was to secure revenues for the king that might otherwise have been overlooked. Often a person's property and valuables were seized upon his sudden or violent death. Revenues were exacted from felons, who forfeited all properties to the crown after they died. Likewise for suicides, which were considered to be acts against God.

It was therefore up to the coroner to investigate all unnatural deaths and to determine cause. But there was a problem. Coroners were not physicians. They were noblemen loyal to and appointed by the crown—more tax collectors than medical experts. And since it was an honor to serve the king, coroners were not even paid.

The office of coroner was exported early on to the English colonies in America and quickly achieved a status comparable to that of dogcatcher. Not only was a medical degree unnecessary, but the only requirement for the job was that a candidate not be an ex-convict. Not an auspicious beginning; and not surprisingly, the profession attracted corrupt and incompetent political hacks. By the end of the Civil War coroners were paid $11.50 per corpse to certify death, and a body might make several rounds before finally being interred. For about fifty bucks homicides would be conveniently certified as natural deaths. Through the coroner's office murderers bought their freedom—cheaply.

Perhaps the situation reached an all-time low when the coroners for Manhattan and Brooklyn each rowed into the East River and began swinging oars at one another in a battle for a floating body. Such spectacles prompted New York City's mayor, John Purroy Mitchell, to replace the office of coroner with that of a medical examiner in 1914. Candidates for the new position had to be physicians as well as pathologists. More important, they had to be trained and experienced in performing *postmortems* (autopsies).

New York was not the first to abolish the office of coroner in favor of an ME. In 1877 the commonwealth of Massachusetts started the trend toward licensed MEs—a trend that continues as states upgrade their systems of investigating unnatural death. Yet even today many states still use coroners who are not forensic pathologists, trained to examine those who have died suddenly, unnaturally, and often violently. The hospital pathologist who examined President John F. Kennedy upon his assassination was well trained in studying disease. Yet he had *never* autopsied a gunshot death before. And his inexperience showed. Unaware that a tracheotomy had been performed on the president that incorporated the exit wound of the bullet, the pathologist was unable to determine how or where the bullet left the body. When x-rays revealed no bullet within President Kennedy, he assumed it had fallen out of the same hole it entered. But bullets don't do that. Additionally, he placed the bullet wound four inches lower than it actually was. According to Michael M. Baden, M.D., and former chief ME for New York City, the postmortem investigation into President Kennedy's death was one of the biggest botch jobs in the history of forensic science. (*Point of information*: President Kennedy's brain, removed during autopsy, was stolen, probably by a memorabilia collector. So was Albert Einstein's.)

It is a very unfortunate fact that forensic pathology is not taught in any medical school. There are no internships or residencies in this "stepchild" of the medical profession. Any doctor interested in forensic pathology must, upon graduation from medical school, train as an apprentice in an ME's office. And there is so much to learn about unnatural death and dying. It is a science that incorporates the wisdom of many different scientific disciplines.

Let's take a closer look at a typical investigation of unnatural death performed by a medical examiner.

Time of Death

One of the first things to be ascertained at the scene of a presumed homicide (or suicide) is time of death. Easier said than done. Medical examiners employ several methods, all of which rely on changes that occur naturally to a body after it dies. One such change, called *algor mortis*, refers to a decrease in body temperature with time. The rule of thumb is 1°F to 1.5°F drop in body temperature for each hour after death. Of course factors such as ambient temperature and size of the body (obese people cool more slowly) will affect the rate of cooling. For this reason the air temperature is recorded along with a rectal-thermometer reading of the body. Algor mortis is most accurate when performed within a few hours after death.

Livor mortis, also called *hypostasis*, is yet another yardstick by which to calculate time of death. It means, literally, the "color of death" and refers to a postmortem reddening of the body. The reddening, however, is not uniform. When the heart stops beating and blood is no longer in motion, gravity takes over, forcing red cells to settle out of the blood. An hour or two after death lividity appears on the lower surfaces of the body—those not in contact with a hard surface. The discoloration first appears as bluish-pink patches. By six to eight hours after death the patches have fused into large purplish areas that blanch when pressed with a finger. After ten to twelve hours the red blood cells break down and leak out of the capillaries (the tiniest blood vessels). They infuse the tissues, and the color fixes, or becomes permanent.

Livor mortis, as a biological clock, is quite inaccurate and unreliable. It does, however, serve another function. By comparing postmortem discoloration with the position of the body, it may be possible to determine whether the corpse has been moved following death. Bodies are often removed from the site of a homicide to conceal what actually took place.

Finally there is *rigor mortis*. It is the mortis we are all familiar with, especially if we enjoy a good murder mystery. Immediately upon death the muscles of the body relax completely and the body becomes slack. *Rigor mortis* refers to the subsequent postmortem stiffening of the body. It begins one to two hours after death, being first noticeable in the muscles of the eyelids. Then the jaw and rest of the face become affected. The stiffness slowly spreads throughout the voluntary and involuntary muscles of the neck, chest, and arms and finally to the lower trunk and legs. The entire process takes about twelve hours. After thirty-six hours the body begins to lose its stiffness, once again becoming slack.

These indicators of time of death, though less than precise, were used effectively in the investigation into the death of John Belushi, famous comic and actor. As Dr. Michael Baden explains in *Unnatural Death: Confessions of a Medical Examiner*, Belushi had been on a four-day drug binge with his Canadian friend Cathy Smith. According to Ms. Smith, she went back to a hotel with Belushi and at 3:30 A.M. gave him his last drug injection—a speedball combination of cocaine and heroin. At 10:15 A.M. she left him alive and sleeping. By 12:30 P.M. he was found dead by his exercise instructor. Was Ms. Smith telling the truth? Let's find out from our friends algor, livor, and rigor.

Rigor Mortis

When Belushi's body was discovered, it was experiencing some stiffness around the jaw, as evidenced by the difficulty Emergency Medical Services (EMS) had inserting a breathing tube into his mouth. This degree of rigor places death approximately one to two hours earlier. Since he was found at 12:30 P.M., death occurred between 10:30 and 11:30 A.M.

Livor Mortis

At about 4:30 the coroner took photographs of Belushi. He was found lying on his back, which showed a distinct purplish color that blanched when pressed (not yet fixed). This was consistent with a time of death six to eight hours earlier. By simple arithmetic we arrive at a time of death of 8:30 to 10:30 A.M.

Algor Mortis

At 4:30 P.M. the coroner also took Belushi's temperature, which was 95°F. If body temperature drops 1°F to 1.5°F per hour, he died at 1:00 P.M. or later. But he was found dead at 12:30 P.M., with rigor already setting in. Clearly the algor calculations were off. Two extenuating circumstances could account for a temperature that should have been considerably lower. First, Belushi was overweight. Second, he was wired on cocaine, which elevates body temperature. Although normal body temperature is 98.6°F, Belushi's could easily have started out at over 100°F. Taking everything into consideration, Belushi could have died at 10:30 but no earlier.

Let's see what the time-of-death indicators tell us:

rigor: 10:30......11:30 A.M.
livor: 8:30 A.M......10:30
algor: 10:30................12:30 P.M.

The overlapping time of 10:30 A.M. was assumed to be the time of death, give or take an hour. This meant that Cathy Smith must have given John Belushi an injection later than 3:30 A.M. Although she never admitted to it, Smith was found guilty of involuntary manslaughter, for which she served fifteen months in jail.

A more recent development in time-of-death determination is the potassium–eye fluid test. After death, red blood cells break down, releasing their potassium, which very slowly leaks into the fluid that fills the eyeball, called the *vitreous humor*. The rate of vitreous absorption of potassium is quite reliable and predictable and is unaffected by temperature. (The cornea of the eyeball, a clear outer covering, also turns cloudy about six to eight hours after death.)

In many cases, where bodies have been dead for a week or more, this method of timing death can be invaluable. Removal of the jellylike eye fluid is accomplished with a fine-needled syringe. To keep the eyeball from collapsing, an equal amount of water is injected back into the eyeball.

Sometimes bodies are discovered weeks or even months after death has occurred. The more time that elapses, the harder it becomes to pinpoint exact time of death. A rough estimate, however, can be arrived at by ascertaining the extent of putrefaction, or decay, of the body. Several days after death a greenish discoloration becomes noticeable on the flanks and abdomen. This staining, which results from bacterial breakdown of the blood, slowly spreads over the entire body. Bacteria of putrefaction also produce hydrogen sulfide and other foul-smelling gases. It is usually the stench of these gases that leads to the discovery of a concealed body.

Five to six days after death, gases have caused a swelling of the face, neck, abdomen, and genitals. The skin also starts to blister. Fluids may leak out of the nose, mouth, ears, and vagina as well. Bloating and blistering continues into the second and third weeks, at which point decomposition can cause soft tissues to liquefy. Skin, hair, and nails become loose and easy to pull off. In warm climates a body can be reduced to a skeleton in three to four weeks.

Putrefaction is what normally occurs to a body. It is the natural course of events following death. It can, however, be prevented from happening under unusual circumstances. Cold, for example, will slow down putrefaction. A body dumped in the snow will hardly decompose at all. If, on the other hand, conditions are warm and dry, as in a desert or a boiler room, *mummification* of the corpse may result. In mummification the entire body dries out and becomes leathery and hard. Skin and flesh shrink around the skeleton. Most important, bacterial action is delayed or prevented entirely. Consequently, internal organs and other body structures remain intact for many, many years. Fingerprints can even be taken from a mummified corpse.

Shortly after death (and sometimes while the person is still alive but unconscious) flies will lay their eggs on a dead body. The eyes are a favorite site of egg deposition. About twelve hours later the eggs hatch into maggots. Further stages in the development of the insect are easily identifiable and, over a period of one to two weeks, follow a self-pre-scribed timetable. With the aid of an entomologist, expert on the life cycles of different maggots, a fairly accurate approximation of time of death can be established. And flies are not the only insects attracted to a corpse. In *Cause of Death*, Keith D. Wilson, M.D., presents this brief timetable of possible insect infestation:

10 minutes:	Ten minutes after body is dead flies arrive and lay thousands of eggs in the mouth, nose, and eyes of the corpse.
12 hours:	Eggs hatch and maggots feed on the tissues.
24–36 hours:	Beetles arrive and feast on the dry skin.
48 hours:	Spiders, mites, and millipedes arrive to feed on the bugs that are there.

(*Point of information*: Hair and fingernails, contrary to popular belief, do not continue to grow after a person dies.)

Cause of Death

By now it has become fairly obvious that a coroner or medical examiner, in addition to needing a strong stomach, must be both doctor and sleuth. Establishing time of death is tricky and inaccurate at best. Yet many times it is a "piece of cake" compared to the ME's other major task, that of determining cause of death.

Cause of death is, hopefully, ascertained in a postmortem examination of the body called an *autopsy*. When performed by an ME investigating a suspicious death, it is more accurately termed a *medical-legal autopsy*. The earliest record of an autopsy being performed in a murder case in the United States was in Maryland in 1665. A man named Francis Carpenter was accused of killing his servant. The postmortem examination showed that Carpenter had indeed bludgeoned the poor man's head, causing a skull fracture, brain hemorrhaging, and death. The verdict: the servant had died because he had failed to visit a doctor after the assault! Sometimes justice is not so blind.

Back to the autopsy table. At first the clothing of the deceased is examined and removed, the external features of the body are inspected carefully. Gross injuries such as gunshot wounds, stabbings, and bludgeoning are noted. X-rays may be useful in tracing the path of a bullet or in determining the angle at which a victim was struck or stabbed. Sometimes the cause of death is not as obvious as would at first appear. If a person is lying on the pavement with a massive injury to the back of the head, one would suspect a bludgeoning murder. Not necessarily. Perhaps the person blacked out and fell backward, smashing his head against the concrete. How can the ME tell whether it is murder or an unfortunate accident? Usually an inspection of the brain will clear up the confusion. If a person is struck on the back of the head with a hard, blunt object, bruising will occur to the back of the brain. If, on the other hand, injury is caused by the back of the head striking the pavement, the front of the skull will be driven into the brain, where bruising occurs. Elementary, my dear Watson.

Bruising and hemorrhaging around the neck suggest strangulation. The likelihood of strangulation becomes even stronger if there are pinpoint hemorrhages in the whites of the eyes, the result of burst capillaries. Opening up the neck and finding fractures to the cartilage of the trachea, larynx, or hyoid bone (located just above the larynx) would confirm strangulation.

With regard to strangulation, it should be noted that many alleged hanging suicides, especially of males, are in actuality the result of a phenomenon known as *sexual asphyxia*. When tightened around the neck, a ligature can constrict the carotid artery carrying blood to the brain. This cuts off the brain's oxygen supply—a condition known as *asphyxia*—and quickly results in loss of consciousness. During masturbation, oxygen deprivation can also heighten the pleasure of an orgasm. Many males experiment with this sort of increased sexual gratification during autoeroticism. The trick is to release the ligature after orgasm but before losing consciousness. It is a close call that often goes against the sexual experimenter.

Asphyxia is not always the result of strangulation or suffocation. When people drown, death is due to asphyxia— an inability of the waterlogged lungs to pass oxygen to the blood. If a body found floating in the river has lungs that are not heavy with water, the person was killed first and then dumped into the river.

Most people who die in fires also succumb to asphyxia—the result of smoke inhalation—before they are consumed by the flames. Smoke contains a high concentration of carbon monoxide. It is the gas in car exhaust fumes and a favorite of suicide victims found dead in their garages. Carbon monoxide does its damage by chemically combining with red blood cells, effectively tying them up so they cannot carry oxygen. Medical examiners are familiar with the telltale signs of carbon monoxide poisoning. It turns the blood a bright cherry-red throughout the body, even in the usually purple or bluish veins. Interestingly, carbon monoxide may still be detected in the blood of a corpse that has been dead for as long as six months. Once again, if a fire has been set to deliberately cover up a homicide, a quick examination of the blood and internal organs (which also turn bright cherry-red) will reveal the deception. Dead people do not breathe in carbon monoxide.

When food swallowed the wrong way blocks the trachea, death by asphyxia soon results. The Heimlich maneuver is designed to dislodge such respiratory obstacles. These choking deaths can, however, be confused with heart attacks, since the symptoms are very similar. The French, in fact, refer to such accidental deaths as *café coronaries*. In his book *Unnatural Death: Confessions of a Medical Examiner*, Dr. Baden tells of a time he was called to an expensive fundraising dinner attended by a gathering of noted physicians. A woman guest had suddenly started gasping and then slumped forward into her food. The doctors were certain it was a heart attack and only needed Dr. Baden to sign the certificate of death. He, however, thought it might be a choking death. Right in the restaurant, he grasped a small steak knife and cut into her throat, near the Adam's apple. Sure enough, said Dr. Baden, ". . . a big, bright green broccoli florette was lodged in the airway, blocking it."

A semifinal note on asphyxia: In early-nineteenth-century England, medicine experienced a renaissance, and medical schools began cropping up like weeds. This created a tremendous demand for cadavers. Enter the body snatchers, or "resurrectionists," as they were popularly called. They were, in effect, grave robbers, digging up freshly interred bodies and selling them to medical schools. Grieving families had to hire grave sitters to watch a grave for three days, by which time decomposition was sufficient to render the corpse worthless.

Even with the looting of graves, the demand for bodies increased steadily. Desperate times call for desperate measures, and two enterprising young ghouls, William Burke and William Hare, came up with a ghastly way of procuring cadavers. In a method of suffocation that came to be known as *burking*, one would sit on the victim's chest while the other pinched the nose and covered the mouth. They were finally caught when a medical student saw his girlfriend lying on the autopsy table.

The beauty of burking, or suffocation in general, is that it leaves an extremely clean corpse. There is no wound, no blood clot, no hemorrhaging, no residual poison in the tissues. The person on the autopsy table seems perfectly fine in every respect except one: he or she is dead.

It is this absence of a detectable cause of death that allowed Mary Beth Tinning, over a fourteen-year period, to murder eight (and perhaps nine) of her young children. Three of the deaths were attributed to sudden infant death syndrome, or SIDS. But if anyone had bothered looking, the circumstances ruled out SIDS. To begin with, some of the infants were awake and being held by Ms. Tinning when they died. SIDS babies die in their sleep, never while being held. Furthermore, Ms. Tinning's babies were cyanotic, or blue, from lack of oxygen. SIDS babies don't turn blue. Finally, a SIDS death occurs to roughly one baby in a thousand. The chances of two occurring to a mother are one in a million. The probability of three is astronomical.

Other Tinning baby deaths were attributed to genetic metabolic errors, cardiorespiratory arrest, Reye's syndrome, or acute pulmonary edema. According to Dr. Baden, these are " 'wastebasket diagnoses,' made in the absence of any true finding." One of the infants, who died of genetic errors, was adopted and not even biologically Mary Beth's and her husband's. So much for bad genes.

Eventually a police chief, suspicious of all the "natural" tragedies that had befallen Ms. Tinning, initiated an investigation. Several of the bodies were dug up and examined. Inconsistencies between her stories and the evidence were presented to her, and she confessed to several of the murders.

Back to the autopsy table. In addition to examining for gross wounds, the body is inspected thoroughly for any marks that might indicate cause of death. Of particular interest are tiny puncture wounds that would be consistent with either homicidal injection of a poison or self-administration of drugs.

Swabs of the mouth, vagina, and rectum are taken and

placed in sterile glass containers. This is done whether sexual contact is immediately apparent or not.

After a thorough external examination the torso of the corpse is cut open. Customarily, this is done in one of two ways. A single incision may be made from the neck, along the midline of the chest and abdomen, down to the pubis. A detour is usually taken around the navel, which is composed of tough connective tissue that is difficult to cut through and even more difficult to sew up afterward. More commonly nowadays, however, the incision takes a Y shape, going from each shoulder to the lower tip of the breastbone and then down. When sewn back up and dressed for the funeral, no stitchings are visible: the autopsy's "bikini cut."

At this time body fluids are collected. At least twenty milliliters of clean blood is taken from the jugular vein after the neck has been opened. Through an incision at the top of the urinary bladder, urine is ladled or pipetted out. These fluids, along with any aspirated from the chest and abdominal cavities, are usually put through a battery of tests designed to detect the presence of alcohol, carbon monoxide, glucose, carbon dioxide, and a number of poisons and drugs. (For a more detailed discussion of poisons, see "Forensic Medicine—A Short Course, Part 2.") Other fluids collected through needle and syringe are cerebrospinal fluid and vitreous humor (from the eyeball, remember?).

Before the body is sewn back up, the heart, lungs, esophagus, and trachea are often removed en bloc and then examined separately. The stomach is opened and its contents collected. In cases of suspected poisoning, the entire stomach as well as the small and large intestines may be removed and saved. Ditto for the kidneys and the liver, which naturally filter out many toxins from the blood and stores them. (*Point of information*: About fifty years ago the livers of older women of fine upbringing showed indentations from the constant pressure of whalebone corsets.)

Finally, the head and brain are examined. First, an incision is made over the top of the skull from ear to ear, cutting through the scalp to the bone. Then the scalp is peeled down over the face, and a power saw is used to open the skull. Finally the brain is removed for analysis.

Have you worked up an appetite yet? Good—now let's break for lunch. (Don't order brains—they're too high in cholesterol!)

Throughout the autopsy many photographs of the body, showing its internal and external condition, are taken. It has even become fashionable to videotape the entire proceedings. If all goes well, the postmortem examination will reveal cause of death. As the Latin inscription on the wall of a medical examiner's office proclaims:

Let conversation cease. Let laughter flee. This is the place where death delights to help the living.

Forensic Medicine: A Short Course

Part 2: Toxicology and Criminal Identification

Forensic Toxicology

Roughly one-fourth of all deaths are brought to the attention of the medical examiner. If he or she suspects foul play, a postmortem is performed. Often, however, the cause of death is not readily discernable. For example, hundreds, perhaps thousands of substances are potentially lethal to humans when inhaled, ingested, or injected. It is impossible for an ME, during an autopsy, to determine which substances are actually present in a corpse in sufficient quantities to cause death. This is the responsibility of *forensic toxicologists* and their bag of chemical tricks.

Toxic, or poisonous, substances can be divided broadly into two categories: *poisons* and *drugs*. The distinction between the two is somewhat arbitrary. If a substance is usually taken in sublethal doses for some alleged beneficial effect (getting high, getting to sleep, alleviating pain or discomfort), it is considered a drug. Otherwise it is simply a poison. Drugs are the method of choice when taking one's life, accounting for upward of 90 percent of all suicides. These figures are only rough estimates, however, since, contrary to popular belief, 75 percent of all suicides leave no notes behind, making it difficult to tell if they are not, in actuality, accidental overdoses.

169

Poisons are more often involved in homicides. They account for only a small fraction of today's murders, yet throughout history certain poisons have achieved a degree of notoriety.

Arsenic

Also called *inheritance powder* for obvious reasons, arsenic has long been a favorite of murderers. In the Middle Ages royalty had tasters to test their foods for arsenic. However, the tasters would not necessarily have done much good: arsenic compounds are odorless and tasteless and kill rather slowly. Four to seven hours after ingestion, severe stomach pain and vomiting occur. The poison severely irritates the stomach, and blood may be found in the vomitus. Diarrhea and intestinal pain follow. Death, which may not occur for another day or later, often results from dehydration and salt imbalance.

In many ways arsenic poisoning is like cholera and other severe gastrointestinal infections. For this reason it has often been overlooked as the cause of death. Many poisons mimic natural causes of death, and because of the expense involved, bodies are not subjected routinely to toxicology screens unless foul play is suspected.

Arsenic's easy availability may be one reason for its popularity. Through the eighteenth and early part of the nineteenth century it was found in a multitude of easily available products, such as weed killer and rat poison. One particularly enterprising young fellow in England even extracted enough of the chemical from flypaper to kill a female friend. To prevent the widespread use of arsenic as a food additive for foul deeds, England passed the Arsenic Act in 1851. It required that all arsenic compounds be mixed with a black or blue colorant so its use could not be disguised easily.

A few interesting things about arsenic: First, its presence in the body discourages bacterial decay; bodies saturated with arsenic decompose more slowly. This bacteriocidal property might explain its use in a variety of medicines, most notably *salvarsan*. Also known as 606 (it was the 606th substance tested), salvarsan was the "magic bullet" used to treat syphilis until the discovery of penicillin.

Arsenic can be found in every part of the body of a poison victim. Even a body that has been exhumed after years of burial will test positive for arsenic in the hair and bones.

With exhumation there is yet another concern to the forensic toxicologist. Arsenic is a fairly ubiquitous substance, occurring naturally in many living organisms in minute quantities, including humans. It is also commonly found in soil, sometimes in high concentrations. When an exhumed body is tested for arsenic, a sample of the soil must also be examined to rule out contamination of the body through seepage from the soil.

Cyanide

Cyanide is very different from arsenic. It acts much more quickly, a small dose of fifty milligrams being sufficient to cause death within five minutes. Like carbon monoxide, it inhibits the blood's capacity to absorb oxygen. Also like carbon monoxide, it can be recognized by the peculiar color it produces in the blood and the lividity of the skin—a dark scarlet.

Cyanide has the added distinction of a characteristic odor, the sweet, sickly smell of almonds so beloved of detective story writers. In actuality this feature of cyanide has been somewhat exaggerated. To begin with, the ability to smell cyanide is determined genetically and is found in only 40 percent of the population. If you possess the proper genes, you might note the faint smell of almonds on the breath of a freshly killed body. At autopsy the smell is also apparent (especially when the stomach contents are removed). But cyanide is quickly broken down by the body, and after a few days the almond odor is no longer discernible. Only the strange lividity remains.

Cyanide, in pill form, was the way Hermann Göring, the Nazi war criminal, avoided the hangman's noose. Many spies, such as Francis Gary Powers, the U2 pilot shot down over Russia in 1960, carry a cyanide pellet for their suicidal convenience in the event of capture by the enemy. Powers himself chose to deal with his Russian captors rather than swallow certain death. It worked out well for him. Convicted of espionage in 1962 and sentenced to ten years of imprisonment, he was released after only two years in a spy-exchange deal.

Perhaps the deadliest form cyanide can assume is that of a highly toxic gas. Used by butterfly collectors in their insect-killing jars, it quickly found more insidious application as an instrument of mass murder of humans. Cyanide was the lethal ingredient in *Zyklon B*, the gas used to exterminate millions of Jews in the Nazi death camps. It was also responsible for more than thirty-five hundred deaths in Bhopal, India, in 1984, when Union Carbide had its tragic chemical disaster. When plastics burn, they release toxic cyanide gas, making cyanide poisoning a common cause of death in airplane fires.

Strychnine

Strychnine is a fast-acting poison (though not as fast as cyanide) derived from the berries of a plant indigenous to India. Supposedly the locals eat the berries to give them immunity to snake venom, even the bite of the deadly cobra. But I wouldn't eat too many berries. Strychnine poisoning is one of the more unpleasant ways to die. It is a sort of internal electrocution with nerve impulses going wildly out of control, causing violent muscle spasms. Brian Lane, in his tome *The Encyclopedia of Forensic Science*, paints the following grisly picture of a typical strychnine death:

> . . . the face is drawn into a characteristic grin due to the contraction of the facial muscles. This grin is known as the "risus sardonicus." Following this,

the muscles are violently and spasmodically contracted, the patient being bent and doubled up into all sorts of shapes. At one moment he may be bent double like a bow, resting on his heels and head—a phenomenon known as *opisthotonos*; at the next he may be jerked off the bed through violent contractions of other muscles.

After a brief respite the attacks continue, this time more violent and agonizing than before:

Muscles of the stomach become hard and tense, the face livid, the eyeballs staring and prominent. Still the patient is fully conscious, though often unable to speak owing to a fixture of the jaw by a variety of lockjaw. The pulse becomes so rapid during the spasmodic attacks as to be uncountable.

During one of the convulsions the victim suffocates due to paralysis of the respiratory muscles. Because of the muscle involvement, strychnine poisoning has been mistaken for diseases such as tetanus and epilepsy, although an epileptic seldom remains fully conscious throughout the ordeal. After death the muscles relax, and it is impossible to tell, without a toxicology report, what caused the death.

Barbiturates

Barbiturates are probably the most dangerous of all the depressants, or "downers." So frequent is their involvement in unnatural deaths, especially suicides, that toxicologists routinely test for these drugs at autopsy. As the term *depressant* implies, barbiturates depress the central nervous system (brain and spinal cord), inducing mild euphoria and sleep. Accidental overdosing is not uncommon since as little as twice the prescribed dosage can be lethal. The subject will fall into a deep sleep, eventually slipping into coma. Death comes quietly, when the victim simply stops breathing.

Barbiturates are what did in Marilyn Monroe, although to this day no one knows whether it was murder, suicide, or an accident. The confusion originally arose when barbiturates were found in her blood but not in her stomach, where they are typically present in accidental overdosing. The implications were obvious: a second party had injected Monroe with the fatal drug.

Perhaps not. As Dr. Michael M. Baden explains in his book *Unnatural Death: Confessions of a Medical Examiner*, it is entirely possible for most of the barbiturates to have been absorbed into Ms. Monroe's bloodstream before she died. And if the toxicology lab was not very exacting (tests were not as sophisticated as they are today), it might well have missed small amounts of gastric barbiturates.

Barbiturates are not the only drugs routinely tested for at a medical examiner's autopsy. Alcohol, cocaine, heroin, and a number of other mind-altering substances are on the checklist. Unlike barbiturates, these drugs are almost always involved in accidental overdosing.

Point of information: In 1952 methane replaced illuminating gas (carbon monoxide and hydrogen) in most American homes as a source of heating fuel. Although explosive, methane is not poisonous. So the head-in-the-oven route was no longer available for would-be suicides. Barbiturates, on the other hand, were increasingly available after World War II. Hence they became the method of choice for ending one's life. While barbiturates were the most commonly used drug for suicide in the 1960s and 1970s, antidepressants such as Elavil and Tofranil may be replacing them in popularity in the 1990s.

Cocaine and Heroin

Although both are mind-altering drugs, cocaine and heroin are very different from one another, both in the pleasurable feelings they induce and in the way they kill. Cocaine produces a "rush," a high. The experience is one of agitation and excitement. Deaths from cocaine are uncommon, but when they do occur they are the result of heart arrhythmia—irregular, uncoordinated beating of the heart that does not allow for proper pumping of the blood. It kills quickly, within half an hour. Len Bias, the college basketball star, died such a needless death.

Heroin, on the other hand, depresses the central nervous system. The euphoria it induces is a "down," a relaxed, peaceful feeling. So relaxed does the brain become that, in effect, it falls asleep and stops sending the signal to breathe.

Sometimes heroin is injected in an effort to bring a person down from a cocaine high. Speedballs are a dangerous combination of cocaine and heroin. When John Belushi died, the forensic toxicologist found enough of both drugs in him to kill two people.

Even before the toxicologist's report comes back from the lab, there are certain unmistakable signs of heroin overdosing. When the heart stops pumping, watery fluids seep out of the capillaries and fill the lungs. Froth works its way up the trachea and into the nose and mouth. Similar effects can be seen after a heart attack. As Dr. Baden explains, "Generally speaking, if we find what looks like a drowning on a rooftop, we know it's drugs. If we find it in an elderly person at home with digitalis pills, it's heart failure."

Criminal Identification

More and more, science and the tools of the scientist are being exploited to help solve crimes. Nowhere is this more evident than in criminal identification. The *comparison microscope*, which allows for the simultaneous viewing of two different samples, is a case in point. Bullets, hair follicles, clothing fibers, and dust particles found at the scene of a crime can be compared with those of a suspect using this double-barreled microscope. It has proved invaluable in bringing criminals to justice.

Yet its importance pales in comparison to that of *dactylography*, or *fingerprinting*, as a unique means of identifying criminals. A turn-of-the-century discovery, it has proved to be the greatest single advance in the science of criminal identification.

Before the advent of fingerprinting, there was little the police could do to establish the guilt or innocence of a crime suspect. England had its *Registry of Distinctive Marks*, which was exactly what its title stated. *Anthropometry*, or the *Bertillon* system, named after its French creator, Alphonse Bertillon, had brief success and popularity in Europe during the late 1800s. It was an inexact and cumbersome system of keeping detailed records of the body measurements of all known criminals. The belief was that no two people had the same exact measurements. But criminal justice had no definitive way of placing a suspect at the scene of a crime.

All that changed with dactylography. As an art and a science, it revolutionized forensics. By July 1901, Scotland Yard's Anthropometric Office was replaced by its Fingerprint Branch, and as they say, the rest is history.

Dactylography

The premise of dactylography is that careful analysis of a person's fingerprint will reveal its unique pattern of whorls, arches, and loops. (Fingerprints are unique to each person, even identical twins, and remain unchanged from the sixth month of fetal development until death.) This is true, but in many if not most instances the prints left behind consist only of invisible secretions of the skin—secretions such as oils, amino acids, urea, and salts (much of it found in sweat). These are called *latent prints*, and they require some means of enhancement to make them visible.

Most often fingerprints are enhanced by dusting with a fine powder—black carbon powder for prints on light-colored surfaces and white aluminum powder for those on dark surfaces. In some cases chemicals are used to reveal prints. One reagent colors skin oils an intense violet. Another turns amino acids purple. In many ways modern fingerprint analysis has become a chemical science unto itself.

Whatever enhancement technique is used, a tape or rubber pad is then used to remove or "lift" the fingerprint so it can be taken to the laboratory for analysis. Interestingly, the lifting of fingerprints was a rather recent development and one not adopted by Scotland Yard until 1970.

In the 1960s dactylography entered the computer age. A scanner that measures light reflected from fingerprint images was used to convert that information into digital data that could be stored in computers. Today police systems have sophisticated computer networks capable of making more than sixty thousand fingerprint comparisons per second. And if a fingerprint is weak or damaged, *laser image enhancement* can sharpen and strengthen the print.

Pretty impressive stuff. The best, however, is yet to come.

DNA Profiling

Using the latest tools of molecular genetics, forensic labs have come up with another high-powered method of criminal identification. It is called *DNA fingerprinting* or *DNA profiling. Profiling* is, perhaps, a better term since the technique has absolutely nothing to do with fingers.

DNA is the hereditary material of living things (see "DNA: The Ladder of Life"), and everyone's DNA is different. Each cell in a person's body has a full complement of his or her DNA. Sperm cells and egg cells contain half that amount yet still more than enough to identify a suspect unequivocally.

DNA profiling, which has been around since 1986, involves removing and analyzing the DNA from blood cells, skin cells, sperm cells, or whatever part the perpetrator inadvertently leaves behind. And not much need be left behind, since a procedure called *PCR (polymerase chain reaction)* can churn out millions of copies of DNA from a minuscule amount. One drop of saliva contains enough skin and/or blood cells to undergo DNA profiling. A single hair (the root part, anyway) was sufficient to convict a British

man of rape in 1989. Special enzymes are used to break up
the long strand of DNA into many smaller pieces called
RFLPs. Because everyone's DNA is different, everyone's
DNA will break into pieces of different lengths and composition. When a suspect's DNA fragments are separated out
on a gelatin surface (through a process called *gel electrophoresis*) and compared with DNA found at the crime
site, positive identification can be established . . . at least
with 99.99999 percent accuracy—or better than one in ten
million.

Rape is a crime particularly suited for DNA profiling
since the rapist leaves so much of himself behind in the form
of several hundred million sperm cells. Perhaps in an effort
to foil the police, one rapist is reported to have used a
condom on his victim. Carelessly, however, he left the condom at the scene of the rape.

For a lucky few, DNA profiling has actually been used
to prove their innocence. Since 1990 about a dozen convicted rapists in the United States have been exonerated
using this procedure on their genetic material.

DNA profiling is also an invaluable tool in searching
for lost and missing people. Between 1976 and 1983 about
nine thousand people were arrested and slaughtered by a
brutal military dictatorship in Argentina. These people were
called the *desaparecido*, the disappeared. Most were young
adults, in their twenties and thirties, with young children.
After their deaths the children also vanished. Evidence suggests that many were taken by childless military couples and
passed off as their own. Now the grandparents of these
children are using DNA testing to locate their long-lost
grandsons and granddaughters.

DNA profiling has also been used to settle disputes of
parentage—whether a person is the biological father or
mother of a particular child. Custody and child support
often hang in the balance.

Scent

Bloodhounds have long been used to track down criminals, following the scent they leave at the scene of a crime. Obviously there is something distinctive about the odor of a person. Each one of us, in fact, has a basic genetically determined odor that is unique and cannot be masked by cosmetics or deodorants.

Working under this premise of individual scents, forensic researchers in laboratories around the globe have been trying to duplicate the feat of tracking dogs. And they are meeting with some success. Of particular interest is a recently constructed "electronic nose" consisting of about a dozen sensors. It can distinguish not only between wine and coffee but also among *different* wines and *different* coffees. Admittedly, the human scent is infinitely more complex, but it is an encouraging beginning.

In an interesting study Dr. Barbara Somerville, of the University of Leeds Department of Forensic Medicine, is collecting sweat in odor-trapping resins and then using gas chromatography to separate the sweat into its several hundred components. This should allow for the creation of a unique sweat profile for each individual, which could be kept in computer files much as fingerprints are.

The advantages of scent identification over fingerprint analysis are obvious. A person's scent cannot be wiped off an object with a piece of cloth. It is at the scene of a crime, omnipresent, whether the perpetrator has worn gloves, galoshes, condoms, or anything else.

Forensics fascinate and spellbinds as no other discipline can. It combines the ingenuity and sophisticated technology of science with the passion and high drama of human tragedy. Both the intellect and the emotions are taken on a wild roller-coaster ride as the mysteries of unnatural death are explored. Sadly, however, forensics is a science that concerns itself with far too many people.

Meteor Showers:
A Raining of Rocks

It is early morning—about 3:00 A.M.—on November 17, 1966, in the western United States. The sky is dark and clear and star-studded. Every minute a streak of light crosses the darkness—a shooting star; a meteor. With growing interest you watch and count. Two per minute . . . then five . . . ten. . . . An hour later more than one is flashing across the sky every second. By 4:30 the rate has increased to 20 per second, or *72,000 an hour*. You can no longer count them. They storm across the sky at a dizzying rate. By 5:00 the storm has become a blizzard. More than 150,000 meteors—*40 per second*—are streaking across the sky every hour. The spectacle is a breathtaking display of celestial fireworks—a heavenly fourth of July.

The storm continues for about an hour, then dies down to a drizzle as dawn arrives, washing out the remains of the greatest meteor event of the twentieth century. You lie there, still dazed by what you have witnessed—and wondering why you hadn't heard that this extraordinary event was coming. The lack of warning was not a simple oversight by the news media. The reason you heard nothing beforehand about the fantastic Leonid meteor storm of 1966 was that *no one* knew it was coming. To understand why even the experts in the field of meteor showers could not forecast it, you have to know a little about meteors.

Meteors: Interplanetary Junk

A *meteor* is a piece of rock or metal or a combination that has entered the atmosphere of Earth at high speed and begun to flare up due to the friction it creates with the atmosphere. It is visible as a streak of light or trail in the sky; we know it as a *shooting star*. Before entering Earth's atmosphere the bit of space debris is known as a *meteoroid*. Curiously, most of the meteors we see shooting across the sky are smaller than a grain of sand. They are visible only because they travel at speeds of more than forty miles per second (sixty-four kilometers per second) and burn very brightly. Larger ones may appear as a ball of fire as they arc across the sky. Hence they are called *fireballs*. At times they may burn as brightly as the full moon, some up to *a hundred times brighter*. Occasionally they break up explosively in midair.

Most meteors burn up completely in the atmosphere. However, if one is large enough—fist size or greater—it will strike Earth's surface before being incinerated. Worldwide, about one meteor strikes Earth's surface every two hours. Meteors that have landed are called *meteorites*. Some people collect meteorites, though it is often difficult to distinguish a meteorite from a surface rock of Earth origin. Metallic meteorites (chiefly iron) are easier to identify than stony ones.

With a meteorite striking Earth every two hours, you might wonder if we should either stay indoors or protect ourselves with crash helmets or steel umbrellas. Completely unnecessary. You are much more likely to be struck by lightning than hit on the head by a meteorite. An injury or death from a falling meteor is reported on rare occasions. On October 9, 1992, a car in Peekskill, New York, was struck by a three-pound meteorite. No one was injured, thankfully, but the car was badly damaged.

Showers vs. Storms

On nearly all evenings, if you stare up at a cloudless and moonless sky for an hour or so, you will see shooting stars. These are *sporadic meteors*—random bits of space debris that happen to enter Earth's atmosphere. (About seven detectable sporadic meteors appear every hour under ideal viewing conditions.) About a dozen times a year meteors flash across the sky at a much greater rate. Such events are generally predictable; they occur each year at the same time. These meteors are not random bits of debris but rather a concentrated belt of particles that Earth crosses regularly in its annual trek around the Sun. This increase in meteoric activity is known as a *meteor shower*. (Occasionally Earth encounters an unexpected, nonrecurring swarm of meteors, but here *meteor showers* means the annual or recurrent type.)

Meteor showers usually produce ten or more shooting stars per hour at their greatest activity, or *peak*. If the rate is excessively high (perhaps one per second or more, but there is no specific rate that defines it), the event is called a *meteor storm*.

The Tail End

Meteor showers are caused by comets. In particular they are caused by comets that cross Earth's orbit. A comet is a member of the solar system that moves around the Sun, as the planets do, but in a much more elongated or elliptical path, with the Sun offset far to one end. Comets are made of water ice, frozen gases of ammonia, carbon dioxide, and cooking gas (also called *ices*), and bits of stone, metal, and dust. Fred Whipple, of the Harvard and Smithsonian Observatories, has described them as "dirty snowballs."

When these snowballs, usually no more than several miles across, approach the Sun in their cosmic journey, they heat up. (The journey may take a few years or millions of years.) The ices begin to vaporize. Within about two hundred million miles of the Sun a dust tail forms. This tail leaves behind a trail of particles, a belt of material that hangs around in space and forms the stuff of meteor showers. The particles in the belt drift very slowly along the orbit of the comet and do not change their position much over the years—so that Earth crosses the path of such a belt around the same time every year.

Meteor showers therefore are associated with particular comets. Table 1 lists the nine most active of these showers, along with pertinent viewing information. The two best displays traditionally are the Perseids and the Geminids. (Of course, specific viewing conditions, such as clear skies and the absence of a Moon, dictate the quality of any meteor shower in any given year.)

Table 1
Meteor Showers Worth Seeing
(in order of appearance)

Shower Name	Peak Date*	Activity** (meteors/hr)	Associated Comet
Quadrantid	January 3	110	?
Lyrid	April 22	12	Thatcher
Eta Aquarid	May 5	20	Halley
Delta Aquarid	July 28	35	?
Perseid	August 12	68	Swift-Tuttle
Orionid	October 21	30	Halley
Taurid	November 4	12	Encke
Leonid	November 17	Variable	Tempel-Tuttle
Geminid	December 14	58	?

*Peak date may vary by a day or two from year to year. Specific viewing information is available in the monthly publications *Astronomy* or *Sky and Telescope*.
**Activity may vary significantly from year to year. Averages over many years are recorded.

The associated comet for each shower is not included in the table—most are not well known and of little general interest. Two showers, however, are caused by a very well-known comet, an eight-mile-wide potato-shaped chunk of ice called Halley, which returns to the Sun every seventy-five to seventy-six years. (It did so last in 1986.) These showers are the Eta Aquarids and the Orionids (see Figure 1).

Figure 1
A Crossing of Paths

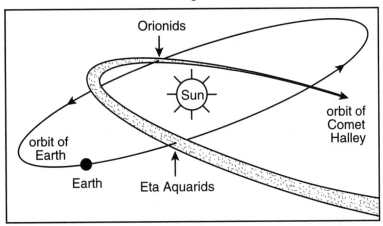

Earth crosses the path of Comet Halley at two separate points, generating meteor showers nearly half a year apart.

The comet associated with the mid-August Perseid shower is also worth mentioning because it returned to the Sun very recently—in late November to early December 1992—after a 130-year loop through space. (Its latest visit before then was during the American Civil War.) Its name is Swift-Tuttle, after comet hunters Lewis Swift and Horace Tuttle. Astronomers believe that on its latest appearance the tail of Swift-Tuttle peppered the Perseid belt with a fresh supply of meteoric dust and that the Perseids of 1993 would be a much more dramatic event than usual, perhaps a storm. (After all, the Perseids of the previous thirty years have been the product of debris deposited *more than one hundred years ago!*) As it turned out, the Perseids put on one of their best shows in recent years—up to 110 meteors per hour, with a high percentage of fireballs—but nothing approaching a storm. There was no storm in 1994 either.

Now you probably have a good idea of why the Leonid storm of November 1966 was not predicted. Very simply, showers are predictble, storms are not. Astronomers cannot predict the out-of-control abundance of meteors that rains across the sky during a meteor storm. The concentration of space junk required for a storm is abnormally high and occupies a very limited and ill-defined region of space within a meteoroid belt.

Taking in a Shower

It is December 14, about three o'clock in the morning. Cold and clear. You stare up into an ink-black sky. Every minute a point of light streaks across it. Some flash briefly; others are more enduring, sweeping halfway across the sky before dying out. You are witnessing the Geminids. But you may also be seeing an occasional random or sporadic shooting star mixed with the shower. How do you tell the difference?

Since meteors that are part of a shower come from a belt of cometary debris occupying a specific region of space, the origin of these meteors—but not sporadics—should be in the same part of the sky. It is the place from which the shower radiates and is called the *radiant*. Meteor showers, in fact, are named for the constellation within which the radiant lies (or, less frequently, a bright star that it is near). Hence the Perseids are named for Perseus, the Rescuer; the Geminids for Gemini, the Twins; the Taurids for Taurus, the Bull; the Orionids for Orion, the Hunter; the Leonids for Leo, the Lion; and so on.

Shooting stars that are part of a meteor shower can appear in almost any portion of the sky. (When viewing a shower, do not look only in the direction of the radiant.) They do, however, give themselves away. The streak, or trail, of a shooting star that is part of a shower will always point in the direction of the radiant (see Figure 2). If played in reverse, the trails will all come together at the radiant. Sporadics do not have a radiant.

Figure 2
Determining the Radiant

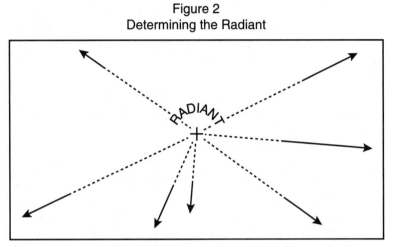

The arrows (solid lines) represent meteor trails. The radiant can be located by extending the trails backward (dashed lines).

When viewing a meteor shower, it's fun to count the number of meteors you see in an hour. Get out a lounge chair or blanket, bundle up if it's cold, lie back, and relax. Don't use binoculars or a telescope; meteor showers are best seen with the naked eye, where your field of view is the widest. It's also fun to plot the meteor trails on a sheet of paper held up to the sky so you can locate the radiant. Some meteor showers also produce shooting stars in various colors or hues. Because you have only two hands and two eyes, seeing and recording everything may be a bit much. Why not "shower" with a friend?

What is the best time of day to view a meteor shower? Of course, it must be at night, with clear skies and preferably no visible Moon. Choose dark surroundings; light pollution can wash out and greatly reduce the spectacle of a meteor shower. Viewing should be as close as possible to the peak. (Often this is not possible, because it is daytime in that part of the world.) Also, it is best to view most showers after midnight and before dawn. At that time, Earth is positioned so that it is rotating into the cometary debris, increasing the velocity of the meteors through its atmosphere. Shooting stars burn more brightly, causing more to be visible and providing a more dramatic display. Before midnight, Earth is rotating away from the meteors, decreasing their overall velocity and brightness.

The great Leonid meteor storm of 1966 was caused by comet Tempel-Tuttle, which leaves a fresh trail of dusty debris every thirty-three or so years. In 1833 and 1866, the debris it left behind caused phenomenal storms. But there was no storm in 1899 or 1933 (though there were showers). This led meteor experts to believe that there was a shift in comet Tempel-Tuttle away from Earth's orbit—and that great Leonid storms were history. Then came 1966—one of the greatest meteor storms ever recorded. The Leonids were not dead after all. Leo had come back to life.

In 1998 or 1999 (it's difficult to predict a comet's orbital period with extreme accuracy), Tempel-Tuttle returns and will once again sprinkle our neighborhood of the heavens with fairy dust. Will the freshly deposited debris create a once-in-a-lifetime event, as it did in 1966, or will it be a dud? No one knows. But on the evening of November 16 after the comet's return, I plan to set my alarm for 3:00 A.M., with a sleeping bag or blanket at the ready. If Leo roars again, I want to be there to see the lights come on.

DNA: The Ladder of Life

Each human cell (except mature red blood cells) contains a grand total of two hundred thousand genes (one hundred thousand pairs), and they hold enough information to fill a stack of textbooks taller than a nineteen-story building. More Nobel Prizes have been awarded to biologists studying the gene than to those engaged in any other scientific endeavor. The study of how living things work is, ultimately, a study of their genes. But what exactly is a gene?

I was afraid you might ask. To Gregor Mendel, an Austrian monk who lived in the mid-nineteenth century, genes (although they were not called genes until twenty-five years after his death) were tiny particulate "factors" of heredity that parents passed down to their offspring. In a patch of monastery garden, Mendel grew and bred many generations of pea plants, following the inheritance of seven different traits (such as plant height, pea color, and pea texture). After eight years of crossing different varieties of pea, Mendel noticed certain patterns of heredity—known as Mendel's Laws—that could best be explained by the existence of these discrete factors of inheritance. He was convinced that a pair of factors governed the expression of each trait. These factors could, but need not, be identical. For example, there was both a yellow and a green version of the factor for pea color. A plant could have two yellow factors, a yellow and a green, or two green factors. In the case of a yellow and a green factor, the yellow was *dominant* and would express its trait over the green. In most differing pairs, one factor was dominant over the other.

Each parent contributed one factor for every trait to its progeny. Whichever of the two factors each parent passed down—through its pollen and egg if a plant or sperm and egg if an animal—determined the genetics of the offspring.

Mendel had originally wanted to work with small lab mammals such as mice or rats instead of plants. The Roman Catholic Church, however, considered animal-breeding experiments unacceptable. It was a good thing for Gregor Mendel that they did, for the color of a mouse's coat is not controlled by a single pair of genes. Most traits are not. A whopping sixty-three gene pairs, working together, determine the mouse's fur color.

Mendel's luck notwithstanding, his experiments and the deductions derived from them were brilliant. He was truly a scientist ahead of his time—so much so that certain shortcomings in his work were inevitable. One such weakness was his total ignorance of the material nature of his heritable factors. What were they? Where were they found in the cell? Gregor Mendel hadn't a clue. How could he know they resided in *chromosomes*, structures that would not be discovered for another twenty years?

Chromosomes are rodlike bodies nestled safely within the nuclei of cells. (Actually, they are rodlike only during cell division and are not visible at other times.) Humans have twenty-three pairs of chromosomes in virtually every body cell. Only mature red blood cells, which have no nuclei, lack chromosomes. Their name derives from an ability to stain darkly with a number of dyes: *chroma* meaning "color" and *soma* meaning "body." To understand how the chromosome's central role in genetics was uncovered, we must turn our attention from the garden variety pea plant to a tiny, harmless creature that loves ripe bananas—the *fruit fly (Drosophila melanogaster)*.

It was this pesky little critter that attracted the attention of Thomas Hunt Morgan and his associates at Columbia University. They began poking and prodding fruit flies with acids, alkalis, radium, and x-rays in an effort to produce new features called *mutations*. The heredity of these mutations could then be followed through many generations (a generation requiring ten days). In this way, Morgan intended either to verify or to disprove Mendel's views, which, at the outset, he did not believe.

For several years Morgan had no luck finding mutations. Finally, one day in May 1910, he looked at one of his fruit flies—all of which had red eyes—and was shocked to see a white-eyed fly peering back at him. This was to be the beginning of a beautiful relationship.

Between 1910 and 1915 Morgan followed dozens of drosophila traits (eye color, wing shape, wing size, body color, etc.) through countless generations. He noticed that two different traits, such as eye color and body color, were *almost* always inherited independently of one another, but sometimes they were not. For example, if a white-eyed, miniature-winged male fruit fly was crossed with a red-eyed, normal-winged female, the two traits of a parent tended to stay together in the offspring. Why?

Enter the chromosome—mysterious inhabitant of the cell's nucleus. For quite some time the nucleus had been suspected of carrying the heredity of an organism since sperm cells, although they are much smaller than eggs, have an equal amount of nuclear material. And to Morgan and his colleagues the newly discovered chromosomes seemed eminently qualified to be the nuclear sites of Mendel's factors (genes). Not only did chromosomes come in pairs, as the genes for traits allegedly did, but the placement of genes on chromosomes offered a simple and logical explanation for the *linkage* of certain traits to one another. To understand why, we must first review some basic biology.

Every human cell has twenty-three pairs of chromosomes. Every fruit fly cell has four pairs—every cell, that is, except the sperm and egg (called *gametes*). When they are produced, from special cells in the testes and ovaries, the chromosome number is halved. This is accomplished by giving each sperm and egg just one chromosome from each pair. Most important, the selection of the chromosome is entirely random. A gamete can receive either chromosome from each chromosome pair.

Such independent assortment of chromosomes has certain consequences for the genes located on them. Genes on different chromosomes will assort randomly, causing their traits to show a random pattern of inheritance.

However, if two or three or even a dozen genes are all situated on the same chromosome, they will be passed as a unit into whichever sperm or egg got that chromosome, and the traits they determine would show a linked pattern of heredity. Figure 1 illustrates this independent assortment and linkage of genes.

Genes a (white eyes) and A (red eyes) are on a different chromosome pair from genes c (hairy body) and C (hairless body); they are not linked. When chromosomes are distributed into sperm and egg cells, all combinations of genes (ac, AC, aC, Ac) are equally likely. Thus, both white-eyed flies and red-eyed flies are equally likely to have hairy or hairless bodies.

Genes a (white eyes) and b (miniature wings) are on the same chromosome; they are linked. The same is true for genes A (red eyes) and B (normal wings). When chromosomes are distributed into sperm and egg cells, linked genes are passed along together. Thus, white eyes and miniature wings are inherited together, as are red eyes and normal wings.

A gene will not, however, faithfully remain with its chromosome mates. In a phenomenon known as *crossing over*, two chromosomes of a pair often swap corresponding pieces before they go their separate ways into a sperm or egg

Figure 1
Independent Assortment of Chromosomes

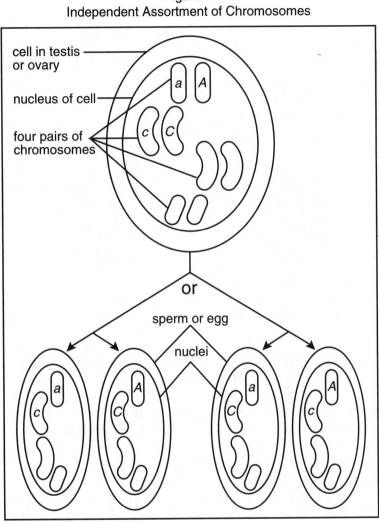

cell. This swapping might separate two linked genes—as shown in Figure 2.

The closer two genes are on a chromosome, the less likely they will be separated by a chromosomal swap. This means that the frequency at which two genes remain linked

Figure 2
Exchange of Chromosome Segments
as a Result of Crossing Over

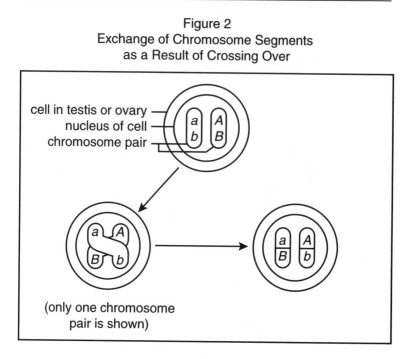

(only one chromosome
pair is shown)

(are inherited together) is a function of their physical closeness on the chromosome. Morgan realized that observing the rate or frequency of linkage could be used to "map" or locate genes at particular sites on a chromosome. By 1915 he had created remarkably detailed linkage maps of the four drosophila chromosomes (only one chromosome of each pair need be mapped). Since Morgan's pioneering work, the phenomenon of crossing over has proved to be an invaluable tool to geneticists in their quest to identify and locate genes.

So genes control specific traits and are located at specific sites on chromosomes. Their arrangement along the chromosome has been compared to pearls on a necklace. Not completely accurate, but good enough for our purposes.

Let's take a closer look at the chromosome. It is a complex structure consisting of different proteins as well as deoxyribonucleic acid (DNA). What part of this chemical mosaic comprises the genetic material?

Unfortunately for genes, the first half of the twentieth century belonged to proteins. They were shown to be amazingly versatile molecules that performed a dazzling array of tasks in living things. And the variation among different protein molecules seemed almost limitless. Small wonder most biologists of the day, including the great double Nobel Prize-winner Linus Pauling, believed proteins were the substance of genes. They were, alas, wrong.

The genetic material turned out to be DNA, a revelation that shocked most geneticists. In the words of the great scientist Max Delbrück: "At that time it was believed that DNA was a stupid substance . . . which wouldn't do anything specific." Oh, how wrong they were, as Oswald Avery clearly demonstrated when, in 1944, he announced the results of several years of research. With coworkers Maclyn McCarty and Colin MacLeod, Avery had purified and identified a substance in bacteria called *transforming principle*. (When scientists call something a *principle*, it usually means they don't know what the heck it is.) This principle, when transferred from a strain of infectious pneumococcus bacteria to a harmless strain, was able to confer virulence on the harmless strain. More important, the transformed bacteria passed this newly acquired trait to their progeny; the principle was inherited. Surprise, surprise, the transforming principle was DNA. In 1952 further support for DNA as the genetic material was provided by Alfred Hershey and Martha Chase at Cold Spring Harbor Laboratory. In an ingenious experiment they showed that when a certain virus infected and reproduced within bacteria, it was the virus's DNA and not its protein that entered the bacteria.

In spite of the evidence, some scientists clung to the notion that proteins were the genetic material. In experiments such as Avery's, they contended, DNA was merely a contaminant. But most of the scientific community accepted Avery's finding. Two scientists who took the finding and ran with it were James Watson and Francis Crick.

DNA was not a molecule new to biologists when Avery

made his disclosure. It had been studied casually ever since
the mid-1870s, when Johann Friedrich Miescher isolated it
from the nuclei of white blood cells and sperm cells. But in
1944 it was suddenly thrust into the hot glare of the scientific spotlight. Labs around the globe began tearing apart
and analyzing the DNA molecule. By 1952 its basic molecular structure had been figured out. The backbone of the
molecule was a long chain of alternating *sugar* and *phosphate*
groups. From this backbone four different types of *nitrogen
bases* dangled, "like so many charms from a bracelet," as
Robert Shapiro described them in his book *The Human
Blueprint*.

What still had to be determined was how DNA func-
tioned as the genetic material. In what way did its structure
code for the myriad events it had to control? And how was
this code passed on to future cell generations?

Watson and Crick answered these questions and, in so
doing, staked their claim to immortality. Both had been
very impressed with the work of Linus Pauling in using a
new technique called *x-ray crystallography* (the study of
x-ray diffraction patterns) to determine the spiraling, or
helical, nature of protein molecules. Watson was convinced
the DNA was also a helix. While studying crystallographs
of DNA he began playing with cutout shapes of the various
DNA subunits, treating them as jigsaw puzzle pieces, trying
to get them to fit into a molecular shape that made sense.
Chemical analysis had shown that two of the four bases,
adenine (A) and thymine (T), were present in equal amounts
in a DNA molecule. The same was true of the remaining two
bases, cytosine (C) and guanine (G). And curiously, the A
and T subunits seemed to fit together nicely as a bonded
pair, and so did the C and G subunits.

Soon x-ray crystallographs from Maurice Wilkins
(with whom Watson and Crick shared the Nobel Prize) and
Rosalind Franklin (with whom they did not) began coming
in depicting DNA as a double-stranded helix. Finally in
1953 all the pieces of the puzzle fell into place. The chain
bracelet with its dangling nitrogen-based charms repre-

sented only half the DNA molecule. It was really two charm bracelets, lined up so that their charms bonded to one another. An A charm of one bracelet always bonded to a T charm of the other, just as a C always bonded to a G; in other words the two strands were complementary. The double chain also had an obvious twist to it. The new picture that emerged resembled a twisted rope ladder more than any jewelry, with the uprights of the ladder being the sugar-phosphate backbone and the rungs representing base pairs.

The uprights of the ladder were of little interest to geneticists; they did not contain the hereditary message handed down from the parent cell. But not so the rungs. These base pairs were like letters of an alphabet, a sequence of base pairs spelling out a mysterious code that told a cell exactly how to go about its business. Unlike the twenty-six-letter alphabet of the English language, the language of the genes was written with only four letters (AT, TA, CG, GC). But a sequence of ten base pairs—a ten-letter word—could have any one of the four base pairs at any of the ten sites. Simple math will tell you that the total possible different ten-base-pair sequences is 4^{10}, or just over a million different combinations. Well, the human genome (all of the DNA in a human cell) is not ten base pairs long. It is a whopping *three billion base pairs*—six billion if we count both chromosomes of each chromosome pair. It's no wonder that there is such diversity in the living world. Figure 3 shows a stretch of double-stranded DNA with an arbitrary sequence of base pairs.

The question now becomes "How can a sequence of bases code for the behavior of a cell?" It does so by controlling the proteins a cell makes. The human body, with its hundred trillion cells, is capable of making at least fifty thousand different kinds of proteins. A protein is actually a long chain of many smaller subunits called *amino acids*. There are twenty different kinds of amino acids, and they must be joined together in the correct order to produce the proper protein.

In the 1960s Marshall Nirenberg of the National Insti-

Figure 3
Model of a Section of DNA
Depicting the Double Helix

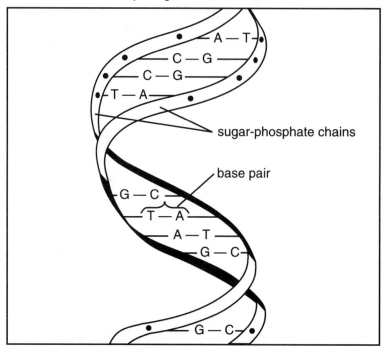

tutes of Health figured out how the sequence of bases in a gene translates into a proper amino acid sequence. It has come to be known as the *triplet code* since three consecutive bases (called a *codon*) on one of the DNA strands specify a particular amino acid. Thus six consecutive bases, or two codons, code for two adjoining amino acids of a protein molecule. In the case of insulin, a stretch of 153 bases is required to direct the assembly of its fifty-one amino acids. Several codons code for no amino acid and, in effect, terminate protein synthesis.

The actual sites of protein assembly are not inside the nucleus; they are tiny particles called *ribosomes* that float around in the cytoplasm. And it is not the DNA that oversees this assembly, for DNA is far too valuable a molecule to go traipsing around the cell. It is kept safely buried in the nucleus, interwoven into the fabric of the chromosomes, like Madame Defarge's secrets. Yet the message must get out to the ribosomes. Enter *mRNA*. It is a molecule that is structurally very similar to DNA, though only single stranded. Strands of mRNA are manufactured along the DNA, using only one of its two strands (called the *coding strand*) as the template. Thus created, mRNA has a base sequence complementary to its DNA template. It is the mRNA—carrying the code from the DNA—that goes to the ribosomes and is directly involved in protein synthesis.

We still haven't answered the question "What's a gene?" Strictly speaking, a gene is a stretch of DNA that codes for a particular protein. There are about one hundred thousand pairs of genes on our twenty-three pairs of chromosomes. Chromosome 1 is our largest and contains about three hundred million base pairs of DNA, divided into roughly ten thousand genes. This is six times as much DNA as in chromosome 21, our smallest one.

Ideally genes should be lined up along the DNA molecule, separated only by codons that end one protein or begin another. No such luck. DNA has proved to be much more complex in structure than that. For starters, there are regions of DNA, called *regulators*, that do not code for proteins but rather turn the bona fide protein-coding genes on and off. The discovery of regulator genes in bacteria earned François Jacob, Jacques-Lucien Monod, and André Lwoff a Nobel Prize in 1965.

Since then, other noncoding elements that control gene activity have been discovered and given colorful names such as *enhancers, promoters, suppressors,* and *silencers.* But it was a 1977 discovery that truly stunned the genetics community. In that year scientists discovered *introns* in the genes of chickens and rabbits. An intron is a stretch of meaningless DNA within a protein-coding gene. Some genes have many introns, which code for no protein and apparently serve only to disrupt the continuity of the gene. Segments of protein-coding DNA within a gene, interrupted by the intervening introns, are called *exons*.

Since 1977 introns have been found in many other animals including humans. In fact the more highly evolved an organism, the more meaningless DNA it seems to possess. Amazingly, more than 95 percent of our own DNA is worthless stretches of introns. A perfect example is the muscular dystrophy gene, which Christopher Wills, in his book *Exons, Introns and Talking Genes*, describes as follows:

> It [the gene] is now known to extend an astonishing 2.5 million bases along the chromosome. The protein coded by the gene is huge, but even so, it accounts for only about eleven thousand bases, half a percent of the total scattered in tiny exons along the gene's length. The rest of the gene is made up of more than sixty-five introns, many of them big enough to conceal dozens of other genes, although nobody yet knows whether they do.

Perhaps introns are leftover bits and pieces of viral infections . . . or bits and pieces of once-functional genes that evolution has cast aside. Whatever their origin, introns were and still are a puzzlement to geneticists. Living cells are supposed to be modules of efficiency, yet the vast majority of our genome is nothing more than junk. And what makes even less sense is that an *entire gene*, introns and all, is transcribed into mRNA. Then, using special enzymes, the cell does a quick editing job, snipping out all the nonsense message and splicing together the remaining exon-coded mRNA before shipping it to the ribosome to make protein. This seems to be a monumental waste of time and energy. To quote Jerold M. Lowenstein, from his December 1992 *Discover* article "Genetic Surprises": "Is this any way to run a genome?"

It certainly is not, which leads some geneticists to speculate that introns may serve some yet undiscovered function. One investigation by Oliver Smithies of the University of Wisconsin seems to indicate that introns are necessary for a gene's expression. When he transferred a particular gene into a cell, it did not express itself (code for a protein) unless several of its introns were included. In some cases introns act as enhancers of genes, turning up their activity from a trickle to a torrent. Yet another possible function of junk DNA was suggested by Natalie Angier in a June 1994 *New York Times* article entitled "Keys Emerge to Mystery of Junk DNA":

> Certain regions of junk may act as reservoirs of change, allowing the DNA to be more easily shuffled, mutated, and rearranged into novel patterns that hasten evolution along [by producing new genes].

DNA is the stuff of life. Following its direction, a single fertilized egg cell becomes a human being or an earthworm or a tree. No event in the natural world is more spectacular or more mind-boggling. And now, for the first time, DNA is beginning to reveal its many incredible secrets to the probings of molecular geneticists. Introns are one of many enigmas to be explored. Regulation of gene expression is certainly another. Why should one group of cells become our eye and another group (which has the *exact same* genome) become our big toe?

Recognizing the importance of a working knowledge of our DNA, scientists and governmental agencies worldwide convened and created the Human Genome Project on October 1, 1990. Its fifteen-year mission is to sequence all of the three billion bases that make up the DNA contained in our chromosomes (only one strand need be sequenced since the other is complementary; that is, a sequence of ATC on one strand would be matched with TAG on the other). Along the way, hopefully, we will learn how and where the bases are arranged into the roughly one hundred thousand functional units called *genes*.

It is a daunting task, however, and one that Dr. Sydney Brenner, an English molecular geneticist, believes can be sped up by studying *fugu*, more commonly known as the puffer. An unusual creature, this fish has a bland, rubbery flesh treasured by sushi eaters. It also has highly poisonous innards, and Japanese chefs must train for two years in its preparation. Still, each year about one hundred people do not survive their platters of raw fugu.

In addition to its toxicity, the puffer has the distinction of being the vertebrate with the smallest known genome— about one-seventh that of humans. (Most mammals, whether human, cat, or mole, have roughly the same three billion bases in their genomes. Inexplicably, some plants have many times this number; for example, wheat has sixteen billion bases and field lilies have one hundred billion bases.) Dr. Brenner is certain that the absent DNA is mostly junk introns and that the puffer has many of the genes we have, in slightly altered form. With much less junk DNA to deal with, it would be far easier to map puffer genes and then use that knowledge to find similar genes in humans. Sounds good to me.

When Worlds Collide

DOOMSDAY ASTEROID: Seven-mile-wide space rock will strike Earth on November 11, say scientists! This is not the first asteroid to threaten Earth—but it might be the last!

So read the July 20, 1993, headline of *Weekly World News*. As it happens, November 11 came and went . . . and we're still here. There was no serious threat to humankind, and the prediction was at best a dramatic overstatement. But it does raise important questions: Are we in any danger from asteroids? Can they and other bodies from space collide with Earth and cause serious if not catastrophic effects?

Things That Go Bump in the Night

Without doubt asteroids can have—and have had—catastrophic effects. It is widely held that an asteroid collision with Earth caused the extinction of the dinosaurs as well as four other periods of major extinction of life on Earth in the past five hundred million years. Such collisions have indeed directed the course of evolution on our planet.

Asteroids are the "minor planets"—heavenly bodies

within the solar system that never accumulated enough mass to become one of the big nine. They are made of either stone (S-type), carbon (C-type), or metal (M-type), and they fall into one of the following three categories based on their location:

1. Asteroid Belt: a region of space between the planets Mars and Jupiter within which asteroids orbit the Sun in near-circular paths. Most asteroids, especially the largest ones, are members of this belt.
2. Trojan Asteroids: two clusters that travel in the same orbit as Jupiter, sixty degrees in front of and sixty degrees behind the Jovian planet. These asteroids are named for heroes of the Trojan War (such as Hector and Nestor).
3. Irregular, or Apollo, Asteroids: a group that has rather elongated orbits that cross inside the orbit of Earth as they round the Sun. The category is named after Apollo, the Greek and Roman god of light, and it was the first asteroid in the group to be discovered. (There are other, lesser-known groups of irregular asteroids that cross the orbit of Mars—the Amors—and that lie mainly inside the orbit of Earth—the Atens.)

Of the three asteroid categories the one that presents a danger to Earth is the Apollos. In crossing Earth's path (which they do, on average, *every few years*), they threaten to collide with Earth. They are predominately of the stony type, with an average diameter of just over half a mile (.8 km). The largest is 24 miles (39 km) across, and the smallest, at latest discovery, is less than 20 feet (6 meters) across. Nearly two hundred Apollos have been discovered so far, with the list growing by several each month. Astronomers estimate that there may be as many as forty-two hundred Earth crossers out there large enough to cause global effects if they ever slammed into us.

The risk of a significant Earth-asteroid collision is not great, but it exists, and over millions or even thousands of years low-probability events manage to occur. Exactly how often have such collisions occurred?

How Often and How Big?

The surface of bodies like Mercury and the Moon represents a lasting geologic record. If you take a look at the Moon through a small telescope or even a pair of binoculars, you will see the pitting and scarring of countless impacts with marauders from space. The face of Earth, however, does not leave such an indelible record. It leaves fewer scars to begin with because its atmosphere burns up many intruders or causes them to explode before they have a chance to land. The scars that do form fade, and the craters smooth over. Earth, you see, unlike Mercury or the Moon, has wind and water that are forever flowing across or raining down on its surface. Over time they act like sandpaper, smoothing irregularities that have been created by collisions with asteriods. The process is called *erosion*.

Even with erosion, geologists have identified about 130 terrestrial impact craters. The list increases by about half a dozen each year. Earth is pockmarked with them. Several examples are the Vredefort Crater in South Africa (6 miles, or 10 km, across), the Ries Structure in Germany (15 miles, or 24 km, across), and in the United States, Arizona's Barringer or Meteor Crater (.75 mile, or 1.2 km, across). The largest-impact crater is in Ontario, Canada. It has a diameter of 124 miles (200 km).

With erosion continually erasing the geologic record, it becomes more difficult to estimate the frequency of Earth-asteroid impacts. Geologists and astronomers, however, do have ballpark figures based on asteroid orbital studies and close observation of surface features of other planets and moons. Their estimates are given in Table 1, which correlates frequency of impact with asteroid size and destructive capability. Keep in mind, however, that these values may vary greatly, depending on the speed and composition of the asteroid as well as the angle at which it enters Earth's atmosphere. On average, asteroids enter Earth's atmosphere at speeds of around ten miles per second (sixteen kilometers per second) or thirty-six thousand miles per hour (fifty-eight thousand kilometers per hour). Most of the smaller ones (150 feet across or less) explode high in the atmosphere, causing little or no damage. This is a key to the destructive capability of an asteroid. "Stones" generally explode more quickly, or higher in the atmosphere, than "metals" do, causing less damage. As Table 1 indicates, an asteroid 150 feet across can be quite destructive . . . if it strikes Earth or explodes near Earth's surface.

It is believed that our Moon may have come about from a glancing blow between Earth and a huge asteroid—one perhaps the size of Mars (about four thousand miles or six thousand four hundred kilometers across), which is more properly called a *planetoid*, or *planet*. The collision would have blasted out material from Earth (and the asteroid) early in its history that later accreted to form the Moon.

Close Ones

Earth has not had a major extraterrestrial collision in nearly a century. Most recent was the Tunguska incident. On June 30, 1908, in a region of central Siberia called Tunguska, a section of forest disappeared. Trees were scorched and knocked down over an area of eight hundred square miles (more than two thousand square kilometers). For many years scientists debated the cause of such destruction. The most logical explanation was a collision with an asteroid. But no impact crater could be found, leading to wild speculation—such as an encounter with a lump of antimatter or a miniature black hole or an unfriendly UFO.

The latest and perhaps best explanation, proposed by Christopher Chyba and his associates at NASA's Ames Research Center in Mountain View, California, is that the perpetrator was a stony asteroid with a diameter of about

Table 1

Size (Diameter)	Frequency	Damage
Under 30 feet (9 meters)	Days to years. Those under one foot in diameter strike Earth daily. They are called meteors.	Varies greatly. The larger ones can cause considerable damage on a local scale.
30–300 feet (9–91 meters)	Years to centuries	Energy release may be greater than a dozen nuclear-bomb explosions.* Can level a city the size of New York in an instant. May strike Earth or explode in atmosphere.
600 feet (about 180 meters)	1,000 years	Very dramatic local effects. Can level an area more than a hundred miles across. Has the energy of about 50 nuclear-bomb explosions.
1,000 feet (about 300 meters)	100,000 years	Forms large craters and ejects vast quantities of dust into the atmosphere. Has the energy of about 10,000 nuclear-bomb explosions, which is greater than the entire world's nuclear arsenal.
3,000 feet or about .6 mile (1 km)	250,000 years	Minimum size for serious global effects. Can disrupt life on Earth and change climate for years or decades, leading to crop failure and starvation. Has the energy of thousands of nuclear-bomb explosions. It is believed that between 1,050 and 4,200 asteroids of this size or larger cross Earth's path.
6 miles (10 km) and larger	millions of years	Believed to have caused extinction of the dinosaurs and 60% of all life forms about 65 million years ago— as well as at least four other mass extinctions. Energy release in millions of nuclear-bomb explosions.

Nuclear-bomb explosion, in this context, means an explosion having the energy release of 1,000,000 tons (1 megaton) of TNT. This is 50 times more powerful than the atomic bombs dropped over Hiroshima and Nagasaki at the end of World War II.

half a football field (150 feet or 46 meters), which heated up in Earth's atmosphere and exploded violently about five miles (eight kilometers) above the surface (hence no impact crater). The shock waves of the explosion leveled hundreds of thousands of trees in the area. If the Tunguska asteroid were iron (M-type) instead of stone, it *would* have struck Earth's surface, leaving a hole the size of Barringer Crater in Arizona and causing much greater damage.

Since 1908 things have been relatively quiet. There was an incident in a town called Revelstoke, in southwestern Canada, on March 31, 1965. Thousands of people saw the explosion of a large meteor (a chunk of interplanetary debris too small to be an asteroid) about twenty miles (thirty-two kilometers) above Earth's surface. It packed the wallop of twenty kilotons of TNT, about equal to the atomic bombs exploded over Hiroshima and Nagasaki during World War II. But, thankfully, it exploded a bit too high for the energy it contained to cause any damage or loss of life. (The Tunguska asteroid had about a thousand times as much energy and exploded four times closer to Earth's surface.)

We have had a number of near-catastrophic close calls in recent years. On June 14, 1968, an Apollo asteroid called Icarus (after a character in Greek mythology who was killed flying too near the Sun on artificial wings) came within 3,700,000 miles (6,000,000 km) of Earth. In March 1989 asteroid 1989 FC came within 690,000 miles (1,100,000 km) of Earth—less than three times the distance to the Moon. It was somewhat smaller in size than Icarus but a thousand times larger than the Tunguska asteroid. Had it struck at even a modest speed of 7 mi/s (11 km/s)—about twice as fast as a speeding bullet—the devastation would have been enormous. It would have whizzed through the atmosphere, crashed, and vaporized in less than one second. The explosion and subsequent shock wave would have leveled everything for more than 155 miles (250 km) around. If it struck over water, it would have created huge tidal waves, flooding coastal areas worldwide, drowning countless numbers of people, and causing untold property damage.

In mid-January 1991 asteroid 1991 BA came within 106,000 miles (170,000 km) of Earth. This is *less than half the distance to the Moon*—very nearly a hit, astronomically speaking. It was not a big asteroid; in fact it was the smallest Earth crosser discovered at the time—a mere 30 feet (9 meters) in diameter. Yet if it had struck or exploded close to Earth traveling at 15 or 20 miles per second, it would have had more explosive force than the Hiroshima or Nagasaki nuclear blasts and would have been able to destroy most of a city.

Comet Collision

Asteroids are not the only heavenly bodies of significant size in the solar system that pose the threat of collision. Comets also can collide with Earth—in particular, comets that cross Earth's orbit in their elliptical journey around the Sun. Such comets were discussed in "Meteor Showers: A Raining of Rocks"; they are the ones that are responsible for Earth's annual meteor showers.

One particular comet that created a bit of a stir recently was Swift-Tuttle, which is responsible for the Perseid meteor shower that occurs every year in mid-August. The comet circles the Sun—and therefore crosses Earth's path—every 130 or so years. Its latest visit was quite recent, in 1992, and astrophysicists predict its return in 2126. Early calculations of its orbital motions prompted the International Astronomical Union to announce, in October of 1992, that there was a one-in-ten-thousand chance that the comet would collide with Earth on August 14, 2126, causing global upheaval, widespread death and destruction, and the end of civilization as we know it. By early December 1992, however, enough additional data had been gathered and studied to rule out even that slim possibility.

Although comets and asteroids are treated as separate entities, there is a connection between the two. Comets are essentially huge snowballs with a small, rocky center. When they visit the inner solar system and round the Sun, the ices vaporize and some of the material is lost. Over thousands or millions of years, the comets may lose all of their ices and become defunct. Their small, stony centers may continue to orbit the Sun. It is believed that at least some of the Apollo Earth crossers are the stony remnants of defunct comets. Others may be asteroids from the belt between Mars and Jupiter that have managed to escape.

Apollo asteroids themselves do not last forever. In crossing the paths of Earth and other planets their orbits are altered, and eventually they crash into one of these planets. One such crash—or crashes—occurred during the week of July 16-22, 1994. The remains of a comet, called Shoemaker-Levy 9, which had broken up into several dozen major pieces (and thousands of minor ones), crashed into the planet Jupiter. It was the first collision of this type ever witnessed by humankind. Although the effects of the encounter were barely visible from Earth, even with a modest-size telescope, and not at all damaging to the Jovian giant, it packed more destructive power by far than all of the nuclear weapons ever detonated on our planet. Thankfully Shoemaker-Levy 9 chose to land on Jupiter rather than Earth.

Eternal Vigil

According to Clark Chapman of the Planetary Sciences Institute in Tucson, Arizona, and David Morrison of NASA's Ames Research Center in California, a major asteroid/comet impact, with global effects, is expected about once in every three hundred thousand years. For a person who lives seventy-five years there would be a one-in-four-thousand chance of experiencing such an impact. Chapman and Morrison also calculated the probability of being killed by an asteroid/comet encounter and compared it to other accidental causes of death in the United States (see Table 2).

The frequency of Earth–asteroid/comet collisions is, of course, statistical. The next major impact could be in five hundred thousand years or in five years. But it does bring

Table 2
Chances of Accidental Death (USA)

Cause of Death	Chances
Motor vehicle accident	1 in 100
Murder	1 in 300
Fire	1 in 800
Firearms accident	1 in 2,500
Electrocution	1 in 5,000
Asteroid/comet impact	**1 in 20,000**
Passenger aircraft crash	1 in 20,000
Flood	1 in 30,000
Tornado	1 in 60,000
Venomous bite or sting	1 in 100,000
Fireworks accident	1 in 1 million

Source: *Nature*, January 6, 1994

home an important realization: in the not-too-distant future (on a geologic time scale), there *will* be a catastrophic collision between Earth and an asteroid or comet. Climate will be altered drastically, and the world's ecological balance will be upset permanently. Much of Earth's plant and animal life will die off—become extinct. Civilization itself may not survive the aftermath. Armed with the awareness of such grave consequences, what are we doing to prevent it from happening?

Not very much, unfortunately. At this point a small handful of people are keeping a "lookout." They include Eugene Shoemaker and his wife, Carolyn (codiscoverers with David Levy of Shoemaker-Levy 9), and a few volunteers. (It was one of these volunteers, Henry Holt, who first spotted asteroid 1989 FC.) Also, a small-scale Spacewatch program, based at the University of Arizona, is continually monitoring the skies. In 1990 the Spacewatch team installed a very sensitive electronic detector called a *charge-coupled-device* (CCD) on a thirty-six-inch telescope at the Kitt Peak National Observatory in Arizona. It is capable of detecting objects much too faint to appear on photographic film. Asteroids as small as sixteen feet (five meters) across have been observed with the CCD-enhanced instrument. (At this size they are approaching the dimension of meteoroids. The size boundary is vague.)

As of June 1993 Spacewatch had identified twelve Earth-crossing asteroids smaller than half a football field in diameter yet still capable of doing significant damage. Their observations also indicate that Earth is struck by several objects *each year* with the explosive energy of a Nagasaki/Hiroshima nuclear blast. Secret military-satellite data place this number much higher—up to *eighty or so* blasts each year. (They generally go unnoticed because they explode high up in the atmosphere. If they exploded lower down, we'd be in big trouble.)

Many astronomers feel that one electronically sensitive large telescope panning the skies for Earth crossers is not enough. According to Eugene Shoemaker, four such scopes placed around the planet could detect 90 percent of the dangerous asteroids or comets within ten years. (An early warning system called Spaceguard Survey, consisting of six large telescopes monitoring the skies for interplanetary interlopers, was proposed by a group of NASA scientists in January 1992 but never approved.) Once detected, one that was headed for Earth could be detonated with high-energy explosives that would either knock it off course or fragment it. We have the technology to pull it off.

The total cost for an effective ten-year Spacewatch program would be about $20 million. That's $2 million per year. It could be shared among the nations of the world since the problem is indeed a global one. This is a very small price tag—a drop in NASA's budget bucket—for ensuring the survival of civilization and perhaps our species.

Just imagine—if an asteroid/comet early-detection and interception program had been in place sixty-five million years ago (with people around to run it, of course), dinosaurs might yet be roaming Earth.

The Hunt for Killer Genes

Midwives used to lick the foreheads of newborn babies they had just delivered. If the skin tasted salty, there was cause for concern, for experience had shown the infant might soon grow ill and die. As it happens, excessively salty sweat is a symptom of *cystic fibrosis* (CF), a fatal disorder that causes sweat glands to function abnormally and lungs and digestive organs to clog with a thick, sticky mucus. This mucus is a breeding ground for bacteria. To prevent deadly pulmonary infections, such as pneumonia, the chest of a CF sufferer must be pounded for several hours each day. This breaks up and dislodges the mucus. Cystic fibrosis patients must also live on a steady diet of antibiotics. Even with all this care, most CF victims die of lung disease or heart failure before age thirty.

Until the early 1980s, doctors hadn't a clue about the mechanisms by which this deadly disease strikes its victims. The only thing they knew was that CF ran in families. CF is, in fact, the most common lethal inherited disease among Caucasian children and young adults, affecting about one in every 2,500 newborns. Unfortunately, there are many other "genetic disorders"—about four thousand in all. Most genetic diseases are extremely obscure and rare, often with weird-sounding names that reflect their most obvious symptoms. *Happy puppet syndrome* (Angelman's syndrome) is a disorder in which the victim experiences prolonged and inappropriate periods of laughter and jerky, puppetlike movements. In *hereditary startle syndrome* there is uncontrolled falling and rigid, muscular movements after a sudden touch or sound. *Maple syrup urine disease* causes seizures and coma in addition to abnormal urine. *Lesch-Nyhan Syndrome* is marked by compulsive self-mutilation of lips and fingers due to biting. These diseases affect comparatively few people, but the more well-known genetic killers—hemophilia, muscular dystrophy, sickle-cell anemia, and Huntington's chorea, for example—have millions of victims.

Ideally, medical investigation of a genetic disease should turn up a substance that clearly and definitively causes the malady. More times than not, the culprit will be a defective protein. Each type of protein made by our cells, and there are many, is under the direction of a particular gene (see "DNA: The Ladder of Life"). By determining the structure of the faulty protein, geneticists can use the genetic code to decipher the structure of the gene and ultimately locate it on a chromosome.

Once they have accomplished that, the ultimate goal is to replace the faulty gene with a functional one. Treating— and, ideally, curing—hereditary disease at this most fundamental cellular level is called *gene therapy*. That may not seem like impressive progress, but for a medical technique that so recently sounded like science fiction, it represents an enormous stride. Thanks to DNA research, we now have new hope of curing diseases once considered hopeless.

One of the first successes was the case of *sickle-cell anemia*. Sickle-cell is a severely debilitating and often fatal disease that primarily affects people of African heritage. The disease derives its name from a collapsing, or sickling, of the red blood cells, which is clearly visible under an ordinary microscope. A multitude of problems result from this sickling, including anemia, excruciating muscle cramps, high incidence of infection, and eventual damage to many organs.

By 1945 the cause of the sickling was determined to be a faulty blood protein called *hemoglobin*. Hemoglobin is the protein that enables red blood cells to haul huge quantities of oxygen around the body. Unfortunately, something goes wrong with the hemoglobin in sickle-cell anemia, and it causes the molecule to twist in a funny way, distorting the red blood cell.

Just what goes wrong in hemoglobin came as something of a surprise to scientists. Like all proteins, the hemoglobin molecule is made of a long chain of amino acids that are strung together in a very special order. In sickle-cell, one amino acid in a chain of 146 is incorrect. Not much of an error, but enough to destroy the three-dimensional structure of the hemoglobin molecule.

In 1978, using the amino acid sequence of hemoglobin and their knowledge of the genetic code, geneticists tracked down the elusive hemoglobin gene. It was the first time that a gene for a major genetic disorder had been identified. Its discovery led, several years later, to a prenatal test for sickle-cell anemia. Today there are prenatal tests for more than seventy-five hereditary diseases.

As it turned out, the small error in the hemoglobin molecule was caused by an even smaller error in the gene coding for it—one incorrect base out of the 1,720 comprising the gene. Many genetic diseases are caused by such single-base, or "point," mutations. It is why Robert Shapiro, in his book *The Human Blueprint*, referred to genetic disorders as "the terrible typos."

Sadly, most genetic disorders do not bend as easily to the will of the researcher as sickle-cell anemia. Defective proteins, which cause the vast majority of hereditary diseases, are difficult if not impossible to isolate. Thankfully, however, there is another tack scientists can take, and it is made possible by the hereditary nature of these diseases: First search out and locate the defective gene, which is at the root of the problem. Then, working backward, discover what protein the genetic material is responsible for producing.

This "reverse genetics" approach is, of course, not as simple as it sounds. We have a huge amount of DNA in our genes, coding for a huge amount of information. If each letter of our genetic code were typed into a book, it would fill five thousand paperbacks. And to make matters worse, the fine detail of DNA is not visible, not even with the most powerful microscope (which happens to be a scanning tunneling electron microscope). Certainly the individual bases of the DNA molecule (see "DNA: The Ladder of Life") are not discernible. Combing through the human genome (all genetic material) in an effort to find a particular gene is akin to looking for a particular sentence in a small library full of books . . . blindfolded.

So where do we begin? Sometimes the search can be narrowed down to a particular *chromosome*. Chromosomes are wormlike structures in the nuclei of every cell (except red blood cells, which have no nuclei). Humans have twenty-three pairs of them, and they carry our one hundred thousand pairs of genes. Twenty-two pairs have two chromosomes that look exactly alike. They are the *autosomes*. The twenty-third pair, the sex chromosomes, determine gender. Males have two different *sex chromosomes*, called X and Y. Females have two morphologically identical chromosomes, both X.

As luck would have it, the genes for certain traits can be assigned immediately to the X chromosome (the Y chromosome is diminutive and carries genes for very few traits; one of these is hairy ears). These are the sex-linked traits, and they are easy to identify—they occur overwhelmingly in males. In 1911, red-green color blindness was the first sex-linked trait to be assigned to the X chromosome. Any male with an X chromosome that harbors the gene for color blindness will automatically be color blind. Females, on the other hand, are protected. For a woman to be afflicted, both of her X chromosomes must possess the abnormal gene. Baldness is another example of a harmless sex-linked trait.

Hemophilia and muscular dystrophy, also sex-linked, are not so harmless.

Muscular Dystrophy

Muscular dystrophy (MD) is a horrible crippling disease in which the muscles of the body deteriorate. Death occurs usually by age twenty, when the heart muscle or breathing muscles become too weak to do their job. Sex linkage does pin down the gene causing MD to a particular chromosome. But the X chromosome still has 166 million base pairs and is bristling with thousands of genes. Where along the DNA helix does the MD gene lie? The search is comparable to looking for a particular house when, as yet, we have identified only the city it is in.

For MD researchers help came in the person of Bruce Bryer, a foster child from Spokane, Washington. Bruce had been born with a trio of devastating X chromosome disorders, one of which was muscular dystrophy. A simple microscopic inspection of the chromosomes from Bruce's white blood cells revealed that a tiny piece of his X chromosome was missing. The search for the particular house had been narrowed to a dozen city blocks.

To understand how the search continued, we must understand DNA—the stuff of our genes. As discussed in "DNA: The Ladder of Life," DNA is a double-stranded molecule. The two strands are joined along their length by a pairing and bonding of their respective base sequences. But the sequences of the two strands are not random. The order of one determines the order of the other, since an A base on one strand will pair only with a T base on the other and a C only with a G. Thus, if one strand has the sequence ATACCGT, the adjoining (complementary) strand must be TATGGCA.

Happily, this selective nature of base pairing enables scientists to hunt down a killer gene. The bloodhound in the hunt is a tiny bit of single-stranded DNA—usually a couple of dozen bases long—called a *probe*, whose whereabouts on a chromosome has already been determined. Because of its single-stranded nature, the probe will bond to any other single strand of DNA with a complementary base pair sequence. This bonding of two complementary strands is called *hybridization*. When mixed with many different single-stranded DNA fragments from a chromosome being studied (made single by gentle heating), a probe, uncannily and with extraordinary precision, finds and hybridizes with its mate. In a technique called *gel electrophoresis*, an electric current first causes the DNA fragments to move and separate through a slab of gelatin. Then a probe, which has been made radioactive or fluorescent, is washed over the separated DNA fragments. After hybridization, the tagged probe thus can easily be located, and it nails down a particular fragment of DNA to the same locus on a chromosome as the probe (see Figure 1).

With this powerful investigative tool geneticists went searching for the MD gene. By the 1970s many probes from each chromosome existed. They were kept in what is termed *DNA libraries* within bacteria, reproducing along with the bacterial DNA.

Using probes taken from healthy people from the area that was missing from Bruce Bryer's *X* chromosome, geneticists performed hybridization experiments on fragments of DNA from many MD sufferers. If they found a probe that would not hybridize with muscular dystrophy DNA, they would know that this segment of DNA was altered in a muscular dystrophy sufferer. Hopefully the alteration would prove to be a mutation in the MD gene—the gene causing the disease.

Eventually Louis Kunkel and Anthony Monaco of Harvard Medical School isolated one probe that did not hybridize with a number of DNA samples. It was a minute bit of genetic material, 200 bases long, and it was a part—a small part—of the MD gene. By 1985 the entire gene, a huge

Figure 1
Hybridizing and Locating a Radioactive Probe

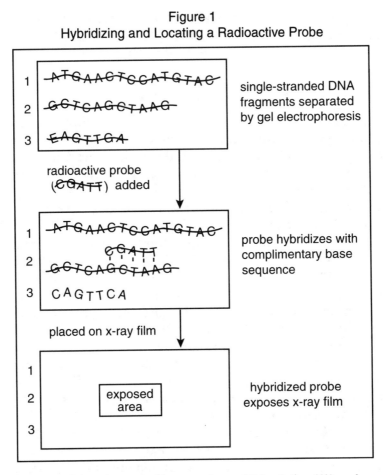

stretch of DNA extending an incredible 2.5 million base pairs, had been uncovered.

The final step was to determine, from the DNA sequence of the gene, the protein that it produced. Once again, easier said than done. Although the gene was uncommonly long, less than 1 percent of it coded for the necessary muscle protein. The rest of the gene was "junk," sequences called *introns*. Nonetheless, by spring of 1988 Kunkel's group had its mystery protein, the one that caused MD.

They called it *dystrophin*. Half a year later dystrophin injections corrected a mouse version of MD. The hunt was over. Call off the hounds . . . I mean the probes.

Cystic Fibrosis

Back to the disease with the salty skin. Researchers investigating cystic fibrosis were not as lucky as those hunting for the MD gene. CF is not sex-linked and therefore cannot be assigned easily to a specific chromosome. Neither was there a Bruce Bryer with a microscopically visible chromosomal abnormality to narrow down the search. So, once again, where do we begin?

We begin by trying to find a linkage between CF and another trait whose gene has already been pinned down to a particular chromosome. This was how Thomas Hunt Morgan mapped dozens of fruit fly genes back in the early 1900s. He looked for pairs of traits that were inherited together, reasoning that they must be joined physically to the same chromosome. Morgan was also able to position genes, relative to one another, on the chromosome by noting how frequently crossing over occurred (see "DNA: The Ladder of Life"). A known gene that is used to nail down an unknown gene is called a *marker*.

Morgan's success rested on his ability to selectively mate thousands of fruit flies over many, many generations. This, however, is not possible with humans, and it posed a major problem for geneticists in quest of defective human genes. The best they could hope for was to find a large inbreeding population that showed a high incidence of the disease—no mean task with diseases that are, in general, quite rare. Once subjects were found, you needed a genetic marker to follow through as many generations of as many family trees as possible, hoping to find a trait that was linked to the disease. Any marker would do, so long as it had already been mapped. The problem was, there simply were not enough well-mapped markers around.

This was the sorry state of affairs CF researchers found themselves in by the early 1980s. They still had no idea which chromosome the CF gene was on. And it was not for want of trying. Then, rather abruptly, things began to change. Human genetics research took a sudden turn for the better. The reason, in a word—make that two words—was *restriction enzymes.* Discovered in 1970, they proved to be of enormous value in the hunt for killer genes.

Restriction enzymes occur naturally in bacteria, where they are used to chop up invading viral DNA. Chopping up DNA is a restriction enzyme's raison d'être. Like molecular scissors, it snips long strands of DNA into smaller pieces. But it does not do so randomly or haphazardly. A particular restriction enzyme will recognize and cut DNA only at a site with a very specific base pair sequence (restriction enzymes snip double-stranded DNA). This selective snipping by restriction enzymes makes them invaluable in the creation of new genetic markers with which to map unknown genes. To understand how these DNA cutters benefit geneticists, let us use an analogy in which a sentence represents a stretch of DNA and its letters the individual bases (or base pairs if we are dealing with double-stranded DNA). The sentence is "Suzy sells seashells." All letters have been capitalized and the spaces between words omitted to make the example clearer:

SUZYSELLSSEASHELLS

Let us now introduce our restriction enzyme, which recognizes only *ELLS* sequences, always cutting the DNA between the *E* and the *L.* After the enzyme has done its thing, the sequence will have been snipped as follows:

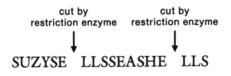

SUZYSE   LLSSEASHE   LLS

Three smaller segments of six, nine, and three bases respectively have been created. Now consider a spelling error, a mutation in the DNA. The *E* in *SELLS* has been erroneously replaced during DNA replication by an *A*. The sentence now reads:

SUZYSALLSSEASHELLS

Introducing the restriction enzyme to this mutated DNA produces the following:

cut by
restriction enzyme
↓

SUZYSALLSSEASHE   LLS

As you can plainly see, only two segments have been produced, one fifteen bases and the other three bases long. A simple typo of a single base pair in the DNA has produced radically different fragments. Any difference or variation in the DNA is called a *polymorphism*. Brown eyes and blue eyes are two polymorphisms at the eye color gene. A, B, and O blood types are three polymorphisms at a particular blood protein gene. The difference in the way a restriction enzyme breaks up two variations of a DNA segment is called a *restriction-fragment-length-polymorphism* (RFLP). Geneticists often refer to them by the shortened name *riflips*, and they can be created all over the place in the human genome.

As more and more restriction enzymes were discovered, riflips began to rain down on geneticists like manna from heaven. And they were a true godsend, harmless alterations in the genome that stood out as signposts along the DNA. They could be followed from generation to generation by a simple application of restriction enzyme to the DNA extracted from white blood cells.

Working with these riflips and a large sampling of CF patients from Toronto's Hospital for Sick Children, in 1985

La-Chee Tsui pinned down the CF gene to chromosome 7. Further linkage studies, with more and better riflips, trapped the gene between two markers 1.6 million base pairs apart. Finally, by the late summer of 1989, Tsui et al. had their gene. Although not nearly as large as the MD gene, it still proved to be quite sizable, with twenty-seven separate protein-coding DNA segments called *exons* coding for a product of 1,480 amino acids. The total length of the gene, interruptions and all, was 250,000 base pairs.

Hunting for the terrible typos has given newfound hope to hundreds of thousands of people who suffer from genetic disorders. In 1990 the gene for elephant man's disease was discovered. By March of 1993, Lou Gehrig's disease (ALS) had been added to the list of diseases with found genes. A month later success was reported in the ten-year search for the genetic cause of Woody Guthrie's killer, Huntington's chorea. September 1994 proved to be a banner month: geneticists announced the identification of five genes associated with insulin-dependent diabetes as well as a gene that causes an inherited form of breast cancer. And there have been other success stories as killer genes continue to be tracked down. But the story does not end with mapping and determining the base sequence of the gene. Ultimately, a cure for the disease must be found. That brings us to *gene therapy*.

Gene Therapy

An ancient Chinese proverb says, "Give a man a fish, and you feed him for a day. Teach a man to fish, and you feed him for a lifetime." Geneticists, in dealing with hereditary diseases, are also looking for the lifetime solution. Give an MD patient an injection of dystrophin or a hemophiliac a shot of clotting factor, and you cure him or her for several days. Repair the genetic-coding error, and you cure that person for a lifetime. Gene therapy is the lifetime cure.

The objective of gene therapy is quite simple and

straightforward: place a correct version of the defective gene into the cells of a patient with a genetic disorder. It was first attempted on a four-year-old girl named Linda, who suffered from an extremely rare form of genetic immune deficiency disease called *adenosine deaminase* (ADA) *deficiency*. Only about fifty persons in the United States have the disease, which is caused by a defective gene's inability to produce a vital enzyme. As a result, immune system cells called *T white blood cells* cannot rid themselves of a certain poison and literally self-destruct. This was the disease suffered by David, the main character in the 1976 movie *The Boy in the Plastic Bubble*, starring John Travolta.

Gene therapy on Linda began on September 14, 1990, but it took years of research to get to that point. The gene causing ADA deficiency had been isolated way back in 1983. Employing restriction enzymes, those marvelous snipping tools used to produce riflips, a normal ADA gene (which produced the vital enzyme) was spliced out of human DNA and into bacterial DNA. As the bacteria reproduced, replicating their DNA at every cell division, thousands of copies of the ADA gene were also replicated. This is called *cloning* the gene.

Once cloning produced sufficient quantities of the ADA gene, a safe, reliable method of delivering the gene to human cells was sought. Enter viruses. A virus is a tiny, infectious particle composed of a protein shell with some DNA (or RNA) inside. What makes it perfect as a vector, or vehicle of transfer, for genes is its mode of infection. Viruses go directly into the cell and use the cell's machinery to make more viral protein and DNA. Certain viruses—the *retroviruses*—even incorporate their own genetic material into that of their host cell's, replicating along with it. What could be more perfect?

With retroviruses in hand, restriction enzymes were used to insert normal ADA genes into the viral genome. Perhaps it should be mentioned that certain genes were also excised from the viral virus, making it incapable of causing illness.

This brings us to Linda and September 14, 1990. At the National Institutes of Health (NIH) in Bethesda, Maryland, Kenneth Culver, R. Michael Blaese, and W. French Anderson hooked Linda up to a machine that removed her white blood cells. The cells were then grown in culture dishes along with the altered retroviruses. For ten days her white blood cells multiplied, hopefully incorporating the normal ADA gene along the way. Finally Linda returned to the NIH, and in a procedure that took less than an hour, her incubated white blood cells were intravenously infused back into her.

Within weeks Linda's immune system showed marked improvement. Her T cell count shot up, and she showed positive reactions to skin tests designed to elicit immune response. She even grew tonsils (which she was born without), which are storage sites for white blood cells. The only downside is that white blood cells live for only a few months, so Linda has had to repeat this procedure every six to eight weeks.

Since then doctors have performed a similar gene therapy on the long-lived bone marrow cells, called *stem cells*, that manufacture white blood cells (see "Our Immune System"). In one immune-deficient newborn, stem cells were retrieved from the infant's umbilical cord, which is particularly rich in these rare cells. If the altered stem cells can produce normal T white blood cells, a lifetime cure to a deadly genetic disorder—ADA deficiency—will have been achieved.

But what about the lethal cystic fibrosis? How has gene therapy fared with this disease? On April 18, 1993, pulmonary specialist Ron Crystal, once again at the National Institutes of Health in Bethesda, snaked a flexible tube called a *bronchoscope* down the throat of a cystic fibrosis patient. A common cold virus was then dripped into the lungs—a virus that harbored a normal version of the cystic fibrosis gene. Hopefully these viral particles will invade respiratory system cells and effect a permanent DNA-level cure. So far the results look promising.

Mars, Our Second Home
Part 1

In the summer of 1976 two spacecraft from the United States landed on Mars: *Viking 1* and *Viking 2*. They scooped up the soil and found it to be composed largely of silicon and iron, chemically combined with oxygen. (In fact the combination of iron and oxygen—as rust—gives the planet its ruddy red appearance.) They sifted the soil and found no evidence of life, not even organic compounds on which life is built. They sniffed the Martian air and found it to be mainly carbon dioxide (95 percent), nitrogen (2.7 percent), and argon (1.6 percent)—an unbreathable mix. They also found it so thin that it would cause our blood to boil—the equivalent of Earth's atmosphere twenty miles (thirty-two kilometers) above sea level. (The top of Mount Everest is less than six miles—ten kilometers—above sea level.) They looked up and found a sky that was salmon pink rather than powder blue. They looked around and found volcanoes the size of Montana and canyons the length of the United States and four times deeper than the Grand Canyon. They searched for liquid water and found not a drop. Their thermometers registered a mean temperature of about −76°F (−60°C), colder than any place on Earth.

We haven't set robot-foot on Martian soil since *Viking*.

The former Soviet Union launched two spacecraft to Mars in 1988, *Phobos 1* and *Phobos 2*, with disappointing results. One failed to reach the Red Planet, and the other failed to land. The United States launched a spacecraft in 1992, *Mars Observer*, which reached Mars and then lost radio contact with Earth. But conditions most certainly have not changed in eighteen years. Mars is a very inhospitable place. Yet it lures us like a Siren. Humankind has had a fascination with the Red Planet from antiquity. It was named for the god of war in Roman mythology—largely because its reddish color was thought to be the spilled blood of Martian warriors. In 1877 an Italian astronomer, Giovanni Schiaparelli, looked through a telescope and saw regular channels—*canali*—crossing its surface. Scientists speculated that they might be the canals or waterways built by an intelligent Martian species—a civilization. This thought was popularized, in particular by the American astronomer Percival Lowell, and persisted well into the twentieth century.

Today we know that these markings are not the construction feat of intelligent beings but natural carvings from rivers of water that flowed over Mars's surface once upon a time. Yes, Mars was much wetter, as well as warmer and gassier, in the distant past. Scientists believe that it might have been warm and wet and gassy enough to support life. If so, what happened to change things on Mars that did not happen on Earth? Can we reverse that process so Mars may once again become a place suitable for life? What technological and ecological miracle would it take to make the fourth planet from the Sun our second home?

What Went Wrong?

Atmosphere is the key. Mars lost its atmosphere over the first billion or two years after its formation, and Earth did not. In both cases it was an atmosphere composed largely of carbon dioxide (CO_2) gas, formed from the eruptions of volcanoes, the vaporizing of meteors either in the atmosphere or on impact, and the liberation of gases trapped in porous rock of the planet's crust—a process called *outgassing*. These natural processes afforded both Earth and Mars a substantial atmosphere in their early history (see "Earth: No Air Mask Necessary"). In turn, the CO_2 atmosphere trapped heat from the Sun in a process known as the *greenhouse effect*. As a result of the greenhouse effect the planets warmed and allowed liquid water to flow and fall as rain. Mars had a bit more difficulty warming up because it is one-and-a-half times farther from the Sun than Earth is, but that was not the main problem. Something else went wrong on Mars.

The answer lies in its size. Mars is considerably smaller than Earth—a bit more than half Earth's diameter, with only one-tenth its mass. Gravity on the surface of Mars is only 38 percent that of Earth. In other words, a one-hundred-pound person on Earth would weigh only thirty-eight pounds on Mars, and a well-struck golf ball would travel half a mile. (Thankfully there are no lakes or ponds on Mars, or my golf ball would find one.) The lesser gravity caused the lighter gas molecules on Mars to escape into space over billions of years.

But gases leaking into space probably were not the main issue. Carbon dioxide is a rather heavy gas molecule and would not have escaped very readily. Size, however, played a role in another regard. Because it is smaller and has less gravity, Mars does not create as powerful an internal pressure or as much subsurface heat as Earth. On Earth pressure and heat cause large sections of crust—called *plates*—to move. This movement is known as *plate tectonics*, and it causes the shifting of continents—or *continental drift*. On Mars, it is believed, no appreciable plate tectonics occur. And that caused Mars to lose its atmosphere.

The connection between plate tectonics and atmosphere is intriguing. Before we explore how it worked on Earth, let's see how it *didn't* work on Mars.

Carbon dioxide gas will dissolve in liquid water to an

extent, forming a weak acid called *carbonic acid*. This substance, in turn, will combine chemically with crustal material, forming carbonates such as limestone. On Mars carbonate formation effectively leached CO_2 out of the atmosphere. Over a period of millions or billions of years, Mars's atmosphere was reduced to less than 1 percent (.8 percent to be exact) of Earth's. Because of its smallness, its center cooled more quickly than Earth's, shutting down volcanic activity that might have replenished the atmosphere somewhat.

Without much of an atmosphere Mars became cold and dry. Liquid water boiled off (low atmospheric pressure will cause water to boil at a low temperature) and then froze directly from the gas. (In the upper atmosphere of Earth snow forms directly from a gas.) Water trapped below the surface froze. Most of the water on Mars today is thought to exist in a solid layer of *permafrost* below its surface and in its polar ice caps (which are mainly frozen CO_2). In fact, at the pressure of Mars's atmosphere water cannot exist as a liquid. There can be no flowing rivers or lakes slapping against rusty shores. As water ice on Mars heats up in the summer months, it passes directly from a solid to a gas. (The process is called *sublimation*.)

On Earth, however, plate tectonics has made all the difference. As Earth's crustal plates moved and shifted, the boundaries of two plates often pressed against each other, with one plate being pushed below the other—an action known as *subduction*. The rock that was forced downward heated gently. This heat caused a release of CO_2 from the carbonates in the rock. The CO_2 gas eventually percolated to the surface, replenishing the atmosphere. Also, volcanic activity continued to add gases to the atmosphere (see Figure 1). As a result Earth stayed warm (greenhouse, remember?), air pressure stayed up, liquid water continued to flow, and life evolved.

Whether or not life had a chance to evolve on Mars is a compelling question. On Earth, life first came about some

Figure 1
Recycling Is the Key

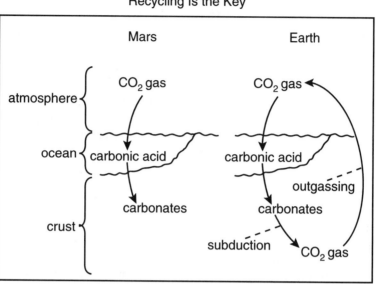

3.8 to 4 billion years ago. Mars, in all likelihood, was still gassy and watery at that time. It is just possible that microbial life *did* evolve on Mars before conditions became too hostile. In that case a record of such life may lie hidden in the permafrost or rock layers below the planet's surface. Less likely is the possibility that life has survived the hostile condition of Mars and *still* exists, holding on tenaciously under the permafrost, where conditions may be warmer and where there may be reservoirs of liquid water in which to splash around (liquid water is a requirement for life; see "The Miracle Molecule"), and where the intense ultraviolet radiation from the Sun, unfiltered by Mars's superthin atmosphere, cannot reach to destroy such life. Conditions in parts of Antarctica are similar to those on Mars in terms of temperature and unavailability of liquid water—yet life of a simple sort survives there.

The only way to answer these questions definitively is to go back to Mars and conduct a comprehensive search for life or fossilized life.

Looking for Life

We have plans to do just that. In fact we should already be there. The NASA spacecraft *Mars Observer* (mentioned earlier) reached the Red Planet on August 24, 1993, and swung into orbit, ready to do its stuff. Then, mysteriously, we lost contact with it and haven't heard from it since.

Russia picked up the space ball from there. The Russians have put together a flotilla of orbiters, landers, and rovers that will visit and probe the planet in a twofold program called *Mars '94/Mars '96*. It is an extensive exploration effort that includes the United States and European Union in a limited capacity.

Mars '94

This mission is scheduled for a September 1994 launch with an ETA of August 1995, eleven months later. Hardware will include two landers and two penetrators, which are spearlike devices that will penetrate Martian soil at over 300 mph (480 km/h). These devices will analyze the soil beneath the surface, record seismic activity, and take weather measurements. The landers will study the soil, possibly for organic compounds necessary for life and for the chemicals that caused Mars to rust. Aerial photos of the planet will also be taken.

Mars '96

This is a more ambitious effort than *Mars '94*. The scheduled launch date is October 1996, with an ETA in September 1997. The mission includes a massive orbiter that will deploy both a rover and a balloon probe. (The balloon probe is a joint French/U.S.-funded project.) The rover will roam the planet's surface, studying the soil and geology over a large area. It may include one or more microrovers that could maneuver into places too small for the main rover. The balloon would rise and fall in the atmosphere as the temperature of the planet heated and cooled from day to night (a nearly twenty-four-hour cycle, as on Earth). It would take weather measurements and get aerial images of the surface. Also, descending from the balloon would be a long rope called the Snake. The end of it would contain instruments to analyze surface and subsurface layers of soil. Full plans for *Mars '96* have not been completed yet. But the efforts should include the search for life on the planet.

U.S. Effort

Most recently (as of February 1994), the United States has entered the field of Mars exploration with a separate program of its own. With President Bill Clinton's endorsement, NASA is planning and has already received initial funding for a modest ten-year program to study the Martian atmosphere, soil, climate, and topography. Starting in November 1996, it will launch two small spacecraft, which will take ten months to reach their destination. One, the *Mars Surveyor*, will orbit the Red Planet over a two-year period, photographing it and collecting weather data. The other craft, called *Pathfinder*, will land a rover on Mars that will study the local terrain for thirty days.

Another set of orbiters and landers is planned for launch in 1998 and again in 2001, scheduled on dates that will take advantage of Mars's closest approach to Earth (which occurs every twenty-six months). Launches beyond 2001 have not yet been determined, but ultimately it is NASA's hope to send out robotic missions that will collect rock, soil, and air samples and return them to Earth.

Yet something is missing—with both the Russian *and* the U.S. program. Neither is attempting to land a human on Mars. Is such a feat possible—a feat accomplished, to this date, only on our Moon? And once there, can we transform Mars into a planet with a breathable atmosphere and water that is potable? Can we turn Mars into a second home?

Believe it or not, serious scientists are giving it serious thought.

Getting There

Before we can think of transforming Mars, we must get there. Balloons hovering in its atmosphere, rovers wheeling across its dusty surface, penetrators stabbing into its rusty rock and soil will not get the job done. We must send *manned* missions to Mars—a journey that is two hundred times longer than a trip to our Moon.

There are several thoughts as to how this will be accomplished. It is estimated that about one hundred tons of fuel would be needed to launch a spaceship to Mars, land it, and then launch and land it back on Earth. (The early colonizers would not yet be going to Mars permanently.) This is too much weight. To reduce the fuel needed, scientists have contemplated constructing the space vehicle in an Earth-orbiting space station, such as *Freedom*, a $30 billion to $40 billion investment scheduled to be completed by 1999. Launching and landing a vehicle from a space station would require much less fuel because it does not have to overcome Earth's gravity.

The problem is that the future for space stations is at best uncertain, especially in today's economic climate. President Clinton has already pressured NASA to design a smaller and less expensive station than the *Freedom* station originally proposed. The Russians have a space station, *Mir*, that has been in operation since February 1986, but whether it could be used to construct an Earth-to-Mars vehicle and whether the Russians are amenable to its use in that capacity are unanswered questions.

A new technology is being explored to reduce the amount of fuel needed to land the vehicle. It is called *aerocapture*. Typically, spacecraft on approaching a planet apply rocket power to slow down and be captured by the planet's gravity, ultimately establishing orbit around the planet. This braking maneuver consumes a huge amount of fuel. In aero-capture the spacecraft would shoot close to but past Mars and out into space. Mars's gravity would curve it into a trajectory and closed orbit around the planet. Without the need to brake the craft, much fuel would be conserved.

Perhaps the best technology is called Mars Direct. It is not really a new technology at all but rather a new way of thinking: using the resources on Mars itself to accomplish the task. It is a way of thinking that we must begin to adopt seriously on all levels if we are to hope to colonize Mars in the future. After all, we cannot bring to Mars oceans of water or a breathable atmosphere in our spaceships.

In the Mars Direct plan, proposed by Robert Zubrin of Martin Marietta Astronomics, the company that built the first Viking lander and *Mars Observer*, spacecraft would be sent to Mars carrying, in addition to the requisite hardware to begin colonization, several tons of liquid hydrogen (H_2). This hydrogen would be made to combine with CO_2 in Mars's atmosphere (which is 95 percent CO_2, remember), forming methane gas (CH_4) and water vapor (H_2O). Methane gas is natural gas. People use it to cook with and to

Figure 2
Gaslight-Era Technology: Burning Cooking Gas

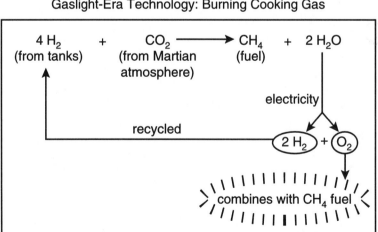

heat their homes. It is a fuel and could be placed into the fuel tanks of the rockets and used to return the astronauts back to Earth. However, methane needs oxygen (O_2) to burn. The oxygen could be gotten by splitting the water that is formed, using electricity. The hydrogen by-product would be recycled into the reaction. (See Figure 2 for the overall chemical process.)

But what happens when we get there? How do we get started changing Mars from a frigid, dry, rusty, nearly airless dustball to a world that is warm and wet and teeming with life? Read on!

Mars, Our Second Home
Part 2

In 1990 President George Bush made a pledge to have human footprints on the rusty soil of Mars by the year 2019. At this point, in the light of a recessional economy, the nearly half-trillion dollars that it would take to make those footprints and the overall trend toward cutting back on space exploration, the pledge of President Bush has a hollow ring. But economies recover and trends reverse. Let us dream a bit and imagine what *might* happen if humans turned their efforts and their technologies toward colonizing the Red Planet and transforming it into a second Earth. (The process of making Mars or any other heavenly body Earthlike is known as *terraforming*, from the Latin *terra*, meaning "earth" or "land.")

Though planetary scientists differ somewhat on specific strategies, the general approaches to terraforming are similar. The following discussion covers the process in three stages, or phases, borrowing largely from the comprehensive study of terraforming Mars done by planetary scientist Christopher McKay and atmospheric scientist Owen Toon of NASA's Ames Research Center and atmospheric scientist James Kasting of Penn State University.

Phase I:
Setting Up Camp (2010 to 2030)

Starting dates may vary greatly, depending on space-research priorities, but from a technological standpoint it could be as early as 2010. This phase would involve landing the first humans on Mars—in space suits, of course. Mars is not an animal-friendly world. Its temperature averages 108°F (60°C) *below* freezing; its atmosphere is almost nonexistent, with a pressure that would cause our eardrums to burst and our blood to boil; its solar radiation would be lethal within minutes of exposure. There is no water to drink or oxygen to breathe. The first colonists would have to live in prefabricated biospheres, or "bubbles," which would house perhaps a dozen people each and would work on the principle of ecological self-sustainment, or the recycling of materials (much as Biosphere II in Arizona did for several years, with limited success).* The colonists would not be poets or politicians . . . or even lawyers! They would be scientists of different sorts—biologists, chemists, geologists, engineers and technicians, medical personnel. . . . Their jobs would be to explore the planet; study its surface and subsurface, atmosphere, climate, radiation, magnetic field; care for those colonists that became ill; and repair those instruments that became inoperable. Communication outside the bubble would be by radio transmission. The cost of this phase is estimated at perhaps $400 billion—far less

---

*Biosphere II began operation on September 26, 1991. It was a sealed, 3.15-acre glass-and-steel enclosure that was supposed to be self-contained and self-sustaining. It was built as a small-scale version of Earth, complete with a tiny ocean, desert, rain forest, grasslands, and nearly three thousand eight hundred species of plants and animals—including four men and four women. Over its two-year life with the original crew, there were serious problems with the maintenance of its atmosphere—additional oxygen had to be pumped in—and with the growth of crops. It was opened several times to bring in supplies.

than Americans would spend on hot dogs, pizza, or cable television in the same period of time.

But Phase I is not terraforming. Living in bubbles and space suits is not creating a new world; it is bringing along the old one. The fun begins in Phase II.

Phase II:
Turning Up the Heat (2030 to 2130)

Very simply, Mars is too cold and has far too little atmosphere to support life. Part of the problem in heating up Mars is that it is 1.52 times farther from the Sun than Earth is and receives only 43 percent as much sunlight. In general Mars is tougher than Earth to heat up and keep warm. But planetary scientists feel that it could be done. The answer, curiously, lies in giving Mars an atmosphere. Look at it this way. When you go to sleep on a cold winter night, you throw a blanket over yourself because it helps to hold in the heat. An atmosphere does the same thing for a planet; it is the greenhouse effect. Mars essentially has no blanket to hold in the heat and keep it warm. In Phase II, Mars is given a blanket, one capable of warming the planet from $-76°F$ ($-60°C$) to above freezing.

A host of methods have been suggested, including the dropping of nuclear bombs into defunct volcanoes, triggering eruptions and the release of large quantities of CO_2 gas and water vapor. The most feasible ones follow.

1. Produce greenhouse gases in chemical factories on Mars and pump them into the atmosphere. Among the most effective greenhouse gases are the chlorofluorocarbons (CFCs)—gases used in refrigerants such as Freon and once used in aerosol propellants, before they were outlawed because on Earth they break down in the upper atmosphere and destroy the important ozone layer. On Mars, however, there is no ozone layer yet—and warming the planet is our chief concern. Astronomers believe that the soil of Mars is rich in the building blocks of CFCs—chlorine, fluorine,

carbon, and hydrogen—and that chemists should have no trouble putting these blocks together to form a whole arsenal of CFCs. According to atmospheric scientist Owen Toon, "Our estimates suggest that it would not take much more than the annual Earthly CFC output—several million tons—to begin warming things up on Mars." The problem is that CFCs are destroyed rather quickly by ultraviolet (UV) radiation from the Sun. A challenge for the terraformers would be to put together CFC molecules that are UV-resistant.

2. Sprinkle or spray a dark material over the polar ice caps. Dark-colored substances absorb infrared radiation, or heat, more effectively than light-colored substances. (That's why you wear dark-colored clothing in winter and pastel shades in summer.) They would warm the polar caps, which are primarily frozen carbon dioxide and water. The warming would cause these ices to vaporize from the solid state (sublimation) and join the CFCs in the atmosphere. Carbon dioxide gas and water vapor are also powerful greenhouse gases. (Water vapor, in fact, is the most significant greenhouse gas in *our* atmosphere.)

The dark-colored materials could be mined from Mars's two moons, Phobos and Deimos, which are among the darkest and most nonreflective bodies in the solar system.

3. Suspend huge Mylar mirrors—many miles across—over the Martian poles to concentrate sunlight on the polar ice caps, vaporizing them and releasing CO_2 gas and water vapor.

As CFCs, carbon dioxide gas, and water vapor build up in the atmosphere, a cycle is set in motion. These gases warm the planet by trapping heat, which in turn causes further vaporization of frozen carbon dioxide and water and in its turn further heating. Also, as the planet warms, carbon dioxide, nitrogen (N_2), and water vapor trapped in the soil and porous rock layers will percolate out and add their measure to the atmosphere. The permafrost—a frozen subsurface sheet of water and carbon dioxide ice—will thaw and vaporize into the atmosphere as well. As a result of all this gassing, Mars will warm significantly. In a few dozen years the temperature could go from $-76°F$ to $-40°F$ ($-40°C$). In fifty to one hundred years it could warm to the melting temperature of water—a Martian heat wave. If, however, the permafrost is deep below the surface—even half a kilometer down—it could take one hundred thousand years instead of one hundred!

The cost of Phase II would be much more expensive than Phase I. We're talking tens of trillions of dollars.

Recent studies of Mars have indicated that, with an increase in temperature and a thawing of the ice caps and permafrost, Mars may become a lot wetter than was once thought—producing ponds, lakes, and even oceans. This is good, of course. Life needs water. But there is a downside. As was mentioned in Part 1, CO_2 has an affinity for water. It dissolves in it and eventually gets tied up with minerals in the crust, forming limestone. On Earth, movement of the crust—plate tectonics—heats up the limestone and releases the CO_2 back into the atmosphere. No such luck on Mars. The terraformers would have to find a way to heat the limestone on a global scale. "Otherwise," says Toon, "the newly created atmosphere would slowly be destroyed." As described in Part 1, astronomers believe this is one of the chief reasons Mars lost its atmosphere in the first place.

Assuming the terraformers can manage to overcome this hurdle—perhaps by mining the limestone and heating it in huge furnaces—things are looking pretty good. The temperature is already at or above freezing. The atmospheric pressure is Earthlike—so that your eardrums would no longer burst and your blood would not boil. The atmosphere would also provide a shield from harmful solar radiation as well as a medium in which sound waves could travel. You would be able to speak with your fellow colonists directly through sound amplifiers in your air mask rather than through radio transmission. People on the planet would number in the tens or hundreds of thousands, but they would still be living in bubbles peppered over the surface of the planet. The reason for air masks and bubbles is simple:

the air of Mars is not yet breathable. And that is the next major miracle to be performed by the terraformers.

Phase III: A Breath of Fresh Air

On Earth the atmosphere became breathable due to a wonderful, green-colored molecule called *chlorophyll*. The magic of chlorophyll is its ability to capture sunlight and use it to split water, releasing the oxygen it contains as O_2 gas. (The process is known as *photosynthesis*.) To be breathable an atmosphere must contain O_2 gas—not too much or the atmosphere would be combustible and not too little or animals would still suffocate. Twenty to thirty percent would be perfect. (Earth's atmosphere is about 21 percent O_2.)

Chlorophyll can be found in all green plants—mosses, grasses, shrubs, trees—as well as microscopic water-dwelling organisms called *algae*. On Earth it took at least hundreds of thousands of years for the atmosphere to become oxygen-rich through photosynthesis. Therein lies the problem. Even if terraformers planted trees and sowed grass seeds over all the Martian land, and filled to the brim with algae every lake and ocean (which would probably not be very salty), it would still take a good one hundred thousand years to make the atmosphere people-friendly. (Plants also need nitrogen. At this point or before, special bacteria would have to be added to the soil and water to convert atmospheric nitrogen into a form that plants could use—namely nitrates.)

Enter the genetic engineers. It would be their task to tinker with the mechanism of photosynthesis, increasing the efficiency of plants and algae to produce O_2 gas. Planetary scientist Christopher McKay feels that this may be the knottiest problem of all. "Plants," he explains, "have been on Earth for billions of years and producing oxygen is what they do for a living. It's possible nature had already optimized this process, already gotten as much oxygen out of plants as is possible."

But perhaps not. Let's be optimistic and assume that plants and algae could be manipulated genetically to "hyperventilate" and that, in addition, chemists and geologists could devise ways of freeing oxygen that is tied up with other elements in Martian rocks. In a hundred years, perhaps, Mars could become breathable. In the meantime the temperature would rise to comfortable levels—possibly 50°F (10°C). A layer of ozone (O_3) would form in the upper atmosphere from the bombardment of normal oxygen molecules with solar radiation. This layer would provide an effective radiation shield for life on the planet. (At this point any lingering greenhouse CFCs would have to be ozone-friendly.) Humans would be able to come out of their bubbles without space suits or air masks. They would look up and see a sky no longer pink from rust dust but Earth blue, with billowing white clouds. It is in this phase of terraforming that Mars would truly become our second home.

The price tag for such homemaking is difficult to estimate. It would certainly be several orders of magnitude greater than that of Phase II.

A Balancing Act

Turning Mars Earthlike is one thing; keeping it that way is another. Planetary scientists believe that Mars was once Earthlike—warm and wet and atmosphere-rich (though not oxygen-rich). The terraformers must work to prevent what once happened on Mars from happening again. Three critical differences between Mars and Earth, already discussed, present problems:

1. Mars's smaller size and mass—with less than two-fifths Earth's surface gravity—means that over a period of time an atmosphere on Mars will leak away much more quickly than it would on Earth. The terraformers must devise ways to replenish this atmosphere.

2. Mars's lack of plate tectonics means the terraformers must devise ways to release the CO_2 gas that gets tied up in

Martian rock, or the atmosphere of Mars will slowly "turn to stone." Heating the rock in furnaces on a global scale might do the trick.

3. Mars's distance from the Sun and relative lack of sunlight make it a tougher planet to keep warm. However, if the terraformers can keep Mars wrapped up in a thick blanket of greenhouse gases, and possibly an orbiting network of solar mirrors to concentrate the attenuated sunlight, the planet should be able to be kept warm enough for human life to exist comfortably. (But bring along your long johns just in case!)

Why Mars?

At closest approach Mars is about two hundred times farther away from us than our Moon is and about twice as far away as is Venus. So why colonize Mars?

Our Moon is a difficult place to terraform because of its small size and mass—much smaller even than Mars. It has about one-sixth the surface gravity of Earth, which means it would be nearly impossible to retain an atmosphere over any extended period. Without an atmosphere there is little hope for the Moon to achieve animal-friendly temperatures or radiation levels or to contain liquid water—or to ever have anything breathable. Presently the Moon is airless and as dry as a bone—actually, much drier.

As for Venus, it is a seething cauldron, with temperatures continually above the boiling point of lead (nearly 900°F, or 480°C), with an atmosphere more than one hundred times denser than our own. The jobs of air-conditioning a planet and reducing its atmosphere are more problematic than heating it up and adding one.

That's about it. With a bit of technological wizardry we've succeeded in terraforming Mars: in turning a dry and barren ball of rock and ices into an incubator for life. But do we have the right to do such a thing, to alter a planet to suit our own purposes? According to Christopher McKay:

> On Earth, the notion of life and the notion of nature are inseparable. But on Mars and in the rest of our solar system, life and nature are two different things. Mars appears to be a dead planet, yet it is undeniably a beautiful, valuable planet. Should we change that natural state?

Do we have that right? McKay answers:

> As far as the ethics of introducing life onto other planets is concerned, I reply that I am prejudiced in favor of life. The *Viking* spacecraft pictures of Mars show a strangely beautiful world, but I think it would be even more beautiful with trees and flowers and little animals scurrying around. I like life.

And so do I!

A Brief History of Time Zones

Ever since humans began walking upright, they have been keeping at least a rough approximation of time. Their timepiece was the Earth itself, rotating on its axis, producing the regular and perpetual alternation of day and night. Not terribly precise, especially over short time intervals, but television had not been invented yet, so accurate, hourly timekeeping was not essential.

As primitive humanity became brainier and civilizations increased in complexity, more exacting methods of telling time were required. By 3000 B.C. Egyptians were already using the sundial to break the day up into smaller, uniform time units. Its operation also depended on Earth's rotation. As Earth spins, the Sun appears to move across the sky, much as a tree appears to move across your field of view when you are on a carousel. (To keep pace with the Sun and have it remain stationary in the sky, a person traveling along the equator would have to move in a westerly direction at 1,038 miles per hour. Over time this value will change slowly, because tidal friction of the oceans against the shores is slowing Earth's rotation by a millisecond a century.) This changing position of the Sun produces changing shadows. At sunrise the Sun creeps over the horizon in the east. A stick called a *gnomon* (which in Greek means "one that knows"), driven straight into the ground, casts a long westward shadow. By noon the shadow has shifted. It is now short and pointing north (in the northern hemisphere). In the afternoon the Sun makes its inexorable journey to the western horizon, once again lengthening the shadow of the gnomon, this time bending it eastward. A slight tilt of the stick allows it to trace a semicircular path on the ground. Add twelve hourly demarcations, and the Egyptians had the first clock.

Inevitably other clocks came along—clocks that did not depend on a sunny day or change the length of an hour with the shifting seasons. There were candle clocks, which marked the passage of time by their rates of burning. Sand clocks and water clocks passed sand or dripped water from an upper to a lower chamber at a rate that marked off the hours. Then, in the mid-1600s, the Dutch physicist Christiaan Huygens made a clock from a swinging pendulum. It was accurate enough to subdivide the hour into minutes. Quite an accomplishment for the seventeenth century. (Today we have atomic clocks based on the oscillations of atoms such as cesium or hydrogen, clocks so precise that they lose less than one second every hundred million years. And yet even more precise clocks are possible with the use of lasers and supercooling.)

Then along came the railroad. It allowed people to travel great distances and connected many far-off towns and cities to one another. Unfortunately, although the time in any one place could be kept quite accurately, every town and city had its own local time. The time in New York City, for example, would be different from that in a neighboring city such as Newark or Southhampton. The reason for these differences is easily understood. Noon (12:00 P.M.) is, by definition, the time when the Sun reaches its highest point in the sky. There are no two points lying east or west of one another for which this time will be exactly the same. When it is noon in New York City, it is just before noon to the west and slightly past noon to the east.

These facts are easily understood but not easily accepted—not if you've got a railroad to run. And the development of the telegraph just exacerbated the problem. By 1884 the situation had become so intolerable that an international convention was held to attempt standardization. What it came up with were time zones.

Time zones are sections of the globe within which all localities have the exact same time. They are defined by lines of longitude, called *meridians*, which run from the north pole, over the surface of Earth, to the south pole. Meridians slice up the planet, like an orange, into twenty-four time zones of equal size. Why twenty-four slices? Because it takes twenty-four hours for Earth to make one complete rotation. Partitioning the globe into twenty-four sections created time zones that would differ from one another by multiples of one hour.

Greenwich, England (near London), because of its

Figure 1
Earth's Time Zones

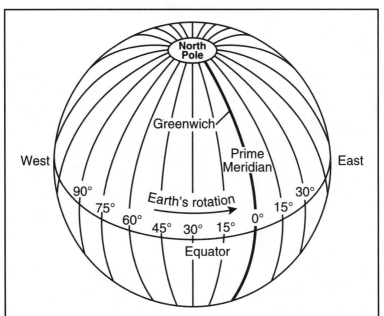

prestigious Royal Observatory, was selected as the starting point for carving out time zones. A line drawn from the north pole to the south pole, passing precisely through Greenwich, was called the *prime meridian* and given the numerical value of zero degrees longitude. Other lines of longitude were drawn every fifteen degrees, effectively dividing the planet into twenty-four equal time zones. Those meridians east of prime meridian were given east longitude designations. Those to the west were given west longitude designations (see Figure 1). At 180 degrees, east met west. The meridian drawn through this point lies on the other side of the planet, directly opposite the prime meridian. It differs in time from the prime meridian by twelve hours. More about this line later.

The prime meridian is not a boundary line separating adjacent time zones. Neither are the other twenty-three lines of longitude measured from it at 15-degree intervals. Rather these meridians are the approximate centers for each of the twenty-four time zones. I say *approximate* because that is exactly what they are. When crossing populated land masses,

Figure 2
Time Zones of the World

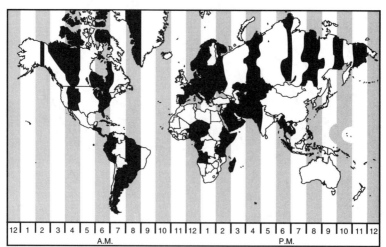

the lines demarcating time zones zigzag, sometimes quite severely, so as to avoid dividing cities, states, and even countries into two different zones. Figures 2 and 3 show how irregular time zones can be as they bend and twist to conform to political and geographic boundaries.

Figure 3
Time Zones of the Continental United States

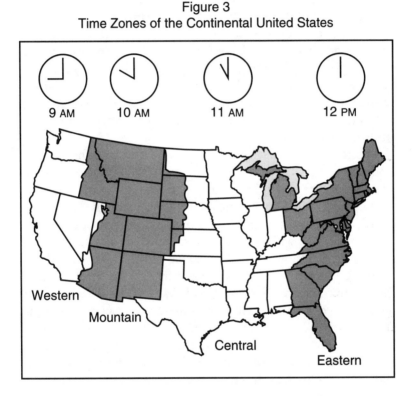

In 1966 the U.S. Congress passed the Uniform Time Act, which established not four but eight standard time zones for the United States and its possessions. In addition to the four zones of the continental United States there are the Atlantic time zone of Puerto Rico (which is an hour earlier than eastern time), the Yukon time zone, Alaska-Hawaii time zone (yes, Alaska and Hawaii are in the same time zone), and Bering time zone of the Aleutian Islands.

Daylight Savings Time

As if things were not already complicated enough, in a whimsical essay the great American scientist/statesman Benjamin Franklin proposed the notion of daylight savings time—of setting clocks an hour ahead of standard time for part of the year. This would create an extra hour of daylight in the evening, at the expense of the morning.

It was not adopted at the time (1784), nor was it taken very seriously. Farmers, in particular, were strongly opposed to daylight savings. Their workday was and still is governed strictly by solar time—when the Sun rises and sets. Moving sunset up from 7:00 P.M. to 8:00 P.M. would mean ending the harvesting an hour later, effectively shortening their after-work evening by an hour.

But by 1915 daylight saving was an idea whose time had come. In that year Germany became the first country to incorporate daylight saving into its calendar. England and the United States quickly followed suit in 1916 and 1918, respectively. By setting clocks an hour ahead of standard time, allowing for an extra hour of daylight when people were up and about, it was hoped that less electricity would be used in homes, and the fuel needed to produce it would be conserved for the World War I effort. Congress, in fact, passed a law putting the entire country on daylight saving time for the duration of the war. This method of energy conservation—permanent daylight saving—was also adopted for about a year, in 1974-75, when oil-producing nations of the Middle East restricted shipments to the United States.

The Uniform Time Act of 1966 put most of the United States on daylight saving for six months a year—from the last Sunday in April until the last Sunday in October. In 1987, as the result of legislation passed by President Ronald Reagan, the start of daylight saving time was moved up from the last Sunday in April to the first Sunday. Interestingly, the U.S. Transportation Department estimates that the earlier starting date is saving more than $28 million in traffic accident costs and preventing more than fifteen hundred injuries and twenty deaths annually. This notwithstanding, several legislatures have voted down daylight saving time. These, which have standard times year-round, include Arizona, Hawaii, Puerto Rico, and parts of Indiana.

Internationally, the situation gets rather confusing. Most of Western Europe is on Daylight Savings Time from the last Sunday in March to the last Sunday in September. England, however, is on daylight saving until the last Sunday in October. Some countries have *double* daylight saving—they are two hours ahead of standard time—while others, notably several Middle Eastern nations, differ from one another by thirty or forty minutes. The clocks of Nepal differ by only ten minutes from those of neighboring India. These countries do not follow the standard hourly demarcations of the time zones. Who ever said telling time was easy?

International Date Line

Between the years 1519 and 1522, Ferdinand Magellan and his skeleton crew of survivors made the first successful around-the-world expedition. When they arrived home, they were baffled to discover that somehow during their travels they had "lost" a day. Their calendar was a day behind friends and family who had stayed at home.

Was their record keeping off? During the many hardships endured throughout the journey, had they forgotten to record a sunrise? No, their record keeping had been precise. What had happened was this: Earth spins from west to east (or in a counterclockwise direction as viewed from above the north pole). By traveling once around the globe in a westerly direction they saw one less sunrise than someone who had stayed home. Viewed another way, had Magellan made his expedition today, he would have had to set his watch back one hour each time he entered a new time zone. By the time his journey was over he would have moved his watch back twenty-four hours—losing a whole day. Conversely, traversing the globe on an easterly course would add a day to your calendar.

But it cannot be Monday for one person and Tuesday for another person in the same location, no matter how many trips around the world one chooses to make. There must be a way of resetting the calendar of a globetrotter to keep it in step with the calendars of those less traveled. That way is the *international date line.*

The international date line is a line of longitude. It is, in fact, the line of longitude that lies approximately 180 degrees from the prime meridian—directly opposite it and on the other side of the world. When one crosses this meridian going west, the calendar moves forward one day to compensate for the lost sunrise. Five P.M. Monday becomes 5:00 P.M. Tuesday. When crossing it going east, the calendar moves back a day.

The 180-degree meridian was chosen as the international date line because it lies primarily over ocean—the Arctic and Pacific to be exact. At one point it does veer to the east to work its way through the Bering Strait, separating Siberia from Alaska. There are several other deviations, as it avoids the Aleutian Islands and other islands of the North and South Pacific (see Figure 4).

Jules Verne made interesting use of the international date line in his classic novel *Around the World in Eighty Days.* The book concerns an English gentleman named Phileas Fogg, who makes a twenty-thousand-pound wager that he can travel completely around the globe in eighty days. The year is 1872, and there is, of course, no air travel to ensure victory, only highly unreliable railroads and less reliable steamboats. Nonetheless, Phileas perseveres, rescuing by elephant a beautiful Indian princess along the way. To make a long story longer, he completes his journey, only to discover that, by his own calculations, he has just missed the deadline and has lost the bet. As luck would have it, however, Mr. Fogg circled the globe in an easterly direction, thereby unwittingly gaining a day as he crossed the interna-

Figure 4
The International Dateline

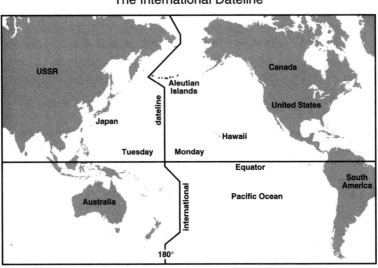

tional date line. He winds up winning the wager as well as the hand of the lovely princess in marriage. Hurray for time zones!

Marriage and wagers notwithstanding, time zones are not always pleasant to deal with. Anyone who has experienced jet lag understands this. Imagine hopping a London-bound plane from New York at 7:00 P.M. Let us assume the flight takes six hours. By the time you touch down at Heathrow airport, your watch and your body tell you it is 1:00 A.M., and visions of sugarplums and your hotel bed dance in your head. But London clocks—five hours ahead of New York's standard time—say 6:00 A.M. A new day is dawning. Do you sleep the next day away or drag yourself around London totally exhausted?

Time zones indeed can be inconvenient. They might even cause a high school student to fail his Earth Science exam (if he were foolish enough to take it before reading this essay). But as long as Earth continues to spin on its axis and the Sun continues to be consumed by thermonuclear fires, time zones will be a necessary part of our lives.

Superconductivity: The Path of No Resistance

Imagine: Magnetically levitated trains traveling at better than 300 mph (483 km/h) on a one-inch cushion of air!

Imagine: Power transmission lines carrying electricity from halfway around the world to light and heat your home without *any* loss of energy.

Imagine: Superpowerful computers that fit in the palm of your hand.

Imagine: Noninvasive medical techniques that do everything from checking your heart function and blood chemistry to detecting ulcers and tumors.

Imagine: High-energy particle accelerators, more powerful than anything we have today, that can help us understand the nature of matter and the creation of the universe.

These are some of the possible products of superconductivity, a technology that promises to catapult us into the middle of the twenty-first century. Long-distance, no-loss power transmission would save huge sums of money, conserve natural resources, and allow nuclear power plants to be built in safe, remote areas of the world. (Remember Chernobyl?) Special superconductor technologies that take advantage of resistance-free circuitry could give us handheld computers with the same power that large supercomputers have today. With improved MRIs (magnetic resonance imagers) and devices called SQUIDS (superconducting quantum interference devices), new horizons in medicine are reachable. Superconducting electromagnets, hundreds of times more powerful than any in commercial use today, can unlock the secrets of the universe.

If we have this technology, why hasn't promise become reality?

One reason is that, at least as far as we currently understand it, superconductors are full of quirks.

What Is Superconductivity?

When you turn on your TV, microwave oven, or air conditioner, the device works because of a phenomenon known as *electricity*. Electricity is the flow of electrons—small, negatively charged particles that exist in all atoms. Electrons flow easily through certain materials, called *conductors*. Most metals are good conductors: copper, aluminum, silver, and gold, to name a few. Mercury, the only metal that is a liquid at room temperature, is a good conductor in both the solid and the liquid state—as are most metals.

But good conductors are not perfect. Electrons in even the best conductors are attracted to or repelled by surrounding atoms or imperfections within the conductor. This causes the electrons to slow down and lose some of their energy, usually in the form of heat. This loss of energy is known as *resistance*, a kind of friction at the atomic level, and it explains why appliances get warm as they run. In fact certain appliances use resistance to fulfill their purpose: electric space heaters, hair dryers, and stoves depend on resistance to produce the desired heat.

Even the best conductors offer resistance. Copper and aluminum wires can cause the loss of as much as 20 percent of electrical energy through resistance. Imagine how much we would save in money and resources if *no* energy were lost! That is the case with a *superconductor*: there is *no* resistance. Electrons flow through the material without any loss of energy. The flow of electricity without resistance is called *superconductivity*.

Heating Up a Cool Customer

With superconductivity not yet widely applied, you might be surprised to learn that we've known that it can work since 1911. In that year a Dutch physicist, Heike Kamerlingh Onnes, discovered that mercury could be made into a super-conductor. The catch was that it had to be cooled down— way down. Room temperature is about 75°F (297 K).* Water freezes at 32°F (273 K). Mercury solidifies at −38°F (234 K). To make it into a superconductor, Kamerlingh Onnes had to cool mercury down to −452°F *(4 K)*. That's just a few degrees above absolute zero, the coldest tempera-ture possible. The air you breathe would turn solid at 4 K. Kamerlingh Onnes was able to get mercury that cold be-cause the technology had been developed to liquefy helium gas. Of all the elements, helium liquefies at the lowest temperature: −452°F. By surrounding mercury with liquid helium, he turned it into a superconductor. It was an amaz-ing discovery and earned the Dutch scientist the Nobel Prize in physics in 1913.

But there were few commercial applications for a super-conductor that had to be cooled down to nearly absolute zero. The cost of cooling to that temperature is prohibitive. Thus began the race to discover, or rather synthesize, mate-

---

*Physicists are partial to the Kelvin (K) scale. In this essay I will use Fahrenheit (F), which we are most familiar with, as well as Kelvin. For a discussion of temperature scales, see "How Cold Is Cold? How Hot Is Hot?"

rials that would superconduct at higher temperatures. The temperature at which superconductivity begins to occur is referred to as the *transition*, or *critical*, *temperature* and is labeled T_c.

The race moved along rather slowly until recently. The materials explored were either pure metals or mixtures of metals, called *alloys*. The best of the lot was an alloy of niobium and germanium, two rare metals. It yielded a T_c of $-418°F$ (23 K) in 1973. This remained the record holder for more than a dozen years, until January 27, 1986. (High-temperature superconductivity, or HTS, generally means above 23 K.) On that day Georg Bednorz and Alex Müller, physicists at IBM's Zurich Research Laboratory, raised the T_c to $-405°F$ (30 K). More important than the temperature breakthrough, however, was the nature of the superconductor. It was not a metal or metal mixture but an *oxide*—a compound containing oxygen. Specifically, it contained the metals barium, lanthanum, and copper along with oxygen, in a specific crystalline arrangement. Metallic oxides of this type are ceramic in nature rather than metallic. The discovery of these ceramic superconductors opened up a whole new area of exploration and earned for Bednorz and Müller the Nobel Prize in physics in 1987.

Carrying the baton in the next phase of the race were physicist Paul Chu and his associates, at the University of Houston. In December 1986 they achieved a T_c of $-387°F$ (40 K)—their first of several world records. By replacing barium with a smaller atom, strontium, they were then able to raise the T_c to $-365°F$ (52 K). Chu worked like an alchemist, mixing different combinations of metals with oxygen in an attempt to find the magic brew: a truly high-temperature superconductor.

It happened on January 20, 1987. Combining *y*ttrium, *b*arium, *c*opper, and *o*xygen (in a compound called YBCO and pronounced "yibco"), Chu got superconductivity at a temperature of $-292°F$ (93 K). So what? It was still a very cold and impractical temperature.

Very cold, yes; impractical, no. In fact this was the discovery of a lifetime. Chu had broken the 77 K barrier—the temperature at which nitrogen liquefies. To quote Robert Hazen, research scientist at the Carnegie Institution's Geophysical Laboratory, where the structure of YBCO was deciphered:

> It was not just another perfect conductor of electricity. This superconductor worked above 77 K, the temperature of cheap, easy-to-handle liquid nitrogen: 77 K is like the sound barrier or the four-minute mile. It is the . . . barrier against which all things cold are measured. Below 77 K any phenomenon . . . is an esoteric curiosity with few practical uses. But anyone can buy liquid nitrogen.* Paul Chu's team had discovered a material that broke the barrier. It could transform superconductivity from an oddity to a day-to-day reality.

Still, the race for high-temperature superconductors continued. Despite the fanfare, −292°F was still quite cold. In 1988 researchers replaced yttrium with thallium, and the T_c rose to −231°F (127 K). Then nothing much happened for five years. Some researchers felt that the temperature limit had just about been reached. They were wrong.

On May 6, 1993, Hans Ott of the Federal Institute of Technology in Zurich reported a superconducting temperature of −220°F (133 K), using a new ingredient in the brew—our old friend mercury. Four and a half months later Paul Chu, the guru of superconductors, got a T_c of −184°F (153 K), using the same material but subjecting it to enormous pressure: 150,000 atmospheres. (The pressure of the air that surrounds us is *one* atmosphere.) Within the next

---

*Liquid helium costs about $4.00 a quart and boils off rapidly. Liquid nitrogen costs about $.10 a quart—*one-fortieth* the price—and lasts about sixty times longer.

month the T_c record was broken several times, and as of August 1994 stands at −164°F (164 K).*

According to Chu, breaking the 150 K barrier was very significant. "For one thing," he said, "we can now cool the material with Freon, using ordinary household air-conditioning technology." That's even cheaper and easier to handle than liquid nitrogen. As for a temperature ceiling, Chu says, "I believe 180 degrees [−135°F] is within sight, although we're not sure how to do it yet." He intends to work his way through the periodic table of elements to find out.

Superconductivity has also come from an entirely different area of materials research: *organic chemistry* (the chemistry of carbon compounds). In 1985 a hollow soccer-ball-shaped molecule was created in the laboratory. Its principal structure consisted of five- and six-sided figures. They were called *buckminsterfullerenes* (*buckyballs* or *fullerenes* for short), after Buckminster Fuller, the architectural engineer who designed the geodesic dome. Buckyballs are miniature geodesic domes—about one forty-billionth of an inch in diameter. They are an interesting class of organic compounds with many fascinating properties—one of them being superconductivity. Although very promising, the T_c for superconductivity among fullerenes is presently far below that of ceramic oxides.

The Meissner Effect

Up to this point superconductivity has been defined as the passage of electricity through a material with no resistance. Another equally important characteristic of superconductors is that they expel a magnetic field. The expulsion is total, and no magnetic field is left inside the superconduc-

---

*An unsubstantiated T_c value of 250 K was reported by a French team in December 1993—that's almost room temperature in Siberia! However, other researchers seriously question the data, describing it as "interesting but not entirely convincing."

tor. This phenomenon was first noticed by Walther Meissner and his graduate student Robert Ochsenfeld in 1933 and is known as the *Meissner effect*. It allows for an interesting classroom demonstration: Place a magnet above a superconductor, and the magnet will remain suspended above it, as if floating in thin air. It is referred to as *magnetic levitation (maglev)* and has an almost magical quality to it.*

But it is not magic. Superconductivity, with all of its accompanying wonders, is very real.

But How Does It Work?

A well-known physicist, Felix Bloch, once said, "The only theorem about superconductivity which can be proved is that any theory of superconductivity is refutable." Gianfranco Vidali, physics professor at Syracuse University, states in his scholarly book *Superconductivity: The Next Revolution?*, "We still don't know how high-temperature superconductivity really works." Even Albert Einstein tried his luck at explaining it, without much success. But we have come a long way.

The answers are complex and embrace a terribly abstract field of physics called *quantum mechanics*. Thankfully, we can gain a general understanding of how it works—or how we *believe* it works—without getting into quantum theory.

In a *normal* conducting wire—a length of copper, for example—electrons in the outermost portion of the atoms are bound very loosely. When connected to a source of electricity such as a battery or an electric generator, these

---

*Anyone interested in such a demonstration can order, through the Institute for Chemical Education at the University of Wisconsin, a kit that includes a one-inch pellet of superconducting compound and a set of magnets. Liquid nitrogen, which is needed to cool the compound down to the critical temperature, is not included. Neither are insulated gloves or safety goggles, which are needed to handle the liquid nitrogen.

electrons are given a push and begin to flow, sliding past atoms that make up the wire. The atoms themselves are arranged in a particular pattern, called a *lattice*. They occupy fixed positions, forming a crystal. However, they do vibrate. At temperatures above absolute zero, all atoms vibrate. (See "How Cold Is Cold? How Hot Is Hot?") The higher the temperature, the more they vibrate. This is one of the main causes of electrical resistance. When atoms vibrate, electrons have difficulty sliding past them. Instead they bump into them. This impedes the flow of electrons and causes them to lose their energy, producing resistance. The lower the temperature, the weaker the vibrations, and the smaller the resistance. However, even at very low temperatures, atoms still vibrate, and there should be *some* resistance. How, then, can superconductors avoid this resistance? Even more puzzling, how can they avoid it at the sweltering temperature of −164°F?

By pairing up. The pairing theory was advanced in 1957 by three American physicists: John Bardeen, who had won the Nobel Prize in physics the year before for inventing the transistor, Leon Cooper, the quantum mechanic from Columbia, and John Schrieffer, the electrical engineer-turned-physicist from MIT. It is called the BCS theory after the first initials of their last names.

The notion of electrons pairing was not a new concept. It had been proposed in the 1930s by physicist Fritz London but failed to gain popularity because it did not adequately explain how electrons in a pair could overcome their natural repulsion for one another. (Blast from the past: Like charges *repel*. All electrons are negatively charged and should repel each other.)

The BCS theory suggested that electrons interacted with the vibrating atoms of the crystal, creating a distortion in the lattice. This distortion acted in a way that pulled the electrons together, in pairs, overcoming their natural repulsion. This electron-lattice interaction has been likened to a heavy ball rolling fast on a soft mattress. To quote Vidali:

If the ball rolls fast enough, the springs will not have time to relax back to the original position immediately after the ball has passed, but instead will take some time to do it. Another ball traveling nearby might encounter the depression and fall into it. We can say that the two balls have "interacted" because they have felt the presence of each other. Since the second ball is pulled to where the first was, we can say that this interaction is attractive.

The interaction causes electrons to pair up in what are known as *Cooper pairs*. And that pairing seems to make all the difference.

John Lagone, in his book *Superconductivity: The New Alchemy*, describes it rather colorfully:

> The electrons in a superconductor may be likened to a platoon of soldiers crossing a field strewn with ruts, holes, and rocks. If they travel singly, they might stumble and fall because of the obstacles. But by pairing off two by two, joining arms, and moving in close formation, they can proceed more smoothly: if one person falls, his partner can hold him up and keep him moving. Since each pair always has another moving steadily in front and one in back, falling down is almost impossible.

For their work in explaining superconductivity, Bardeen, Cooper, and Schrieffer were awarded the Nobel Prize in physics in 1972. It was Bardeen's second Nobel in physics, a feat accomplished by no other individual.

The BCS theory applied remarkably well to the earlier, conventional superconductors, the lower-T_c metals and metallic alloys that were the backbone of superconductivity until the mid-1980s, when the higher-temperature superconductivity explosion started. Then came the age of the new ceramics—the metallic oxides. The BCS theory does not seem to explain fully HTS in these materials, of which copper and oxygen are the principal ingredients. It does appear that Cooper pairs are still significant, but other processes may be at work. Scientists have noted that the amount of oxygen is critical—the more the merrier—and that these atoms connect with the copper, forming chains and planes through which superconductivity current seems to flow. Other metals in the ceramic are also important in that they provide the electrons for the Cooper pairs. No one is exactly sure what's going on. According to Arthur Sleight of DuPont Industries, "There's at least one theory for every theorist."

Promises, Promises

None of the theorists has come up with a solution to the problems that keep superconductivity's promise from being fulfilled. One of those problems has to do with the Meissner effect. Unfortunately, if the magnetic field is strong enough, it can overcome the expulsive force of the superconductor, penetrate it, and destroy its superconductivity. The point at which this occurs is called the *critical magnetic field*. The problem here is that electric currents produce magnetic fields. (It is the reason why huge electromagnets are envisioned in the superconductivity technology of the future.) The situation becomes a catch-22. To carry sufficient current to be useful commercially, superconductors would create magnetic fields that would destroy their own superconductivity. At this point in time the critical magnetic field of superconductors is not high enough to make their use practical on a large scale.

There is some hope, however. Superconductors fall into two categories: Type I and Type II. Type I's are the older variety, the pure metals that superconduct only at very low temperatures. They have low critical magnetic fields. Type II's are the metal mixtures and oxide ceramics that work at higher temperatures. Their critical magnetic fields are considerably higher and offer the possibility of being raised to commercially feasible levels in the not-too-distant future.

But there is another problem. Superconductivity is lost when current flow through the material increases beyond a certain point, called the *critical current density*. Presently, superconductors that operate in the liquid nitrogen range have critical current densities nearly one hundred times too low to be practical commercially.

The situation is indeed frustrating. The potential for sweeping change is there for a technological revolution on the grandest scale. But the problems are there as well. It is like an extended good news/bad news joke. Superconductivity has been put to limited use by the military and by private industry, but it has not yet delivered its promise commercially. Either the technology is not yet there, or the cost is prohibitive. Take maglev trains, for example. We can build them if we want—but at a cost of about $30 million for every mile of track laid down.

Each year the promise comes closer to reality. Better materials are being concocted by the "new alchemists," allowing for higher critical temperatures, higher critical magnetic fields, and higher critical current densities. To this point *nine* Nobel Prizes have been garnered by these new alchemists. How many more will it take before we can travel to work on a cushion of air and turn on a light with electricity that came all the way from a nuclear power plant in Antarctica?

Our Immune System

In 1976 a made-for-television movie, *The Boy in the Plastic Bubble*, introduced us to David, a youngster born without a functioning immune system. He had to live in an absolutely germ-free environment because exposure to any bacterium, virus, or protozoan could cause a fatal infection.

In the 1980s David received a bone marrow transplant from his sister in an attempt to establish an immune system. Tragically, and completely unknown to David's physicians, the Epstein-Barr virus lay hidden within his sister's marrow cells. Although the virus normally causes a not-too-serious disease called *mononucleosis* (the "kissing disease"), in David's defenseless body the virus ran hog-wild, ravaging him with cancer (yes, viruses cause cancer). Within four months he died, tumors riddling his intestine, liver, lungs, and brain.

David died because he lacked an immune system, described by Edward Edelson in his book *The Immune System* as "a collection of tissues, cells, and molecules working together to recognize and attack the small enemies that prowl about the world, looking for ways to make biological profit at the expense of the human body." Chief among these tissues, cells, and molecules is an amazingly diverse

269

array of *white blood cells*. They are the guts of the system. Flowing through the blood as well as the lymphatic vessels of the body, white blood cells collect in the spleen, tonsils, adenoids, appendix, small intestine, and dozens of lymph nodes. There they lie in wait for invading microorganisms, with which they do constant battle. Understanding our immune system is understanding our many different kinds of white blood cells.

The first true insights into where and how white blood cells form came from the horrific atomic bomb blasts at Hiroshima and Nagasaki at the end of World War II. People exposed to high doses of radiation died from internal bleeding and/or infection within ten to fifteen days. This *radiation syndrome*, which apparently destroyed, among other things, the immune system, could be treated successfully by injecting bone marrow cells from a genetically compatible donor. Eventually it was discovered that all types of white blood cells (as well as red) originate from special rare bone marrow cells called *stem cells*.

In 1988, at the Stanford University laboratories of Irving L. Weissman, stem cells were first identified in mice. Since then human stem cells have been isolated and studied intensively. The many types of white blood cells that they spawn can be grouped broadly into several different cell lines, each playing a crucial role in the immune response.

Nonspecific Response

The most primitive and least specialized are the *phagocytes*, from the Greek word meaning "to eat." A phagocyte is a white cell that oozes along like an amoeba. When it encounters a bacterium or virus, it engulfs and devours it. The process is aptly termed *phagocytosis*. *Macrophages*, large amoeboid cells, are one common example of phagocytes. Phagocytes are nonspecific—they will attack and swallow any kind of invading germ.

So will *natural killer* cells, a second line of white blood cells. Unlike phagocytes, however, natural killer cells do not attack free-roaming germs. Their targets are body cells that have gone bad—those that harbor viruses or have turned cancerous. Natural killer cells kill not by engulfing but by punching holes in cell membranes.

Nonspecific immune responses such as these are an important line of defense. They are in fact the only immune system possessed by invertebrates—simple boneless animals such as insects, worms, clams, and starfish. Unfortunately, humans need much more than this. Luckily, they have much more—a third and exceedingly vital class of white cells, the *lymphocytes*.

Figure 1
Derivation of Different White Blood Cells

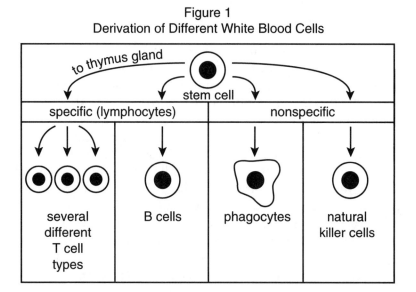

Specific Response

Without question lymphocytes perform the lion's share of defense against microbial attack. They are the infantry, the cavalry, and the air force all rolled up into one. Make that two, for although all lymphocytes look alike under the microscope, there are actually two very distinct lymphocyte populations—the *B lymphocytes* (B cells) and the *T lymphocytes* (T cells). The differing paths these two populations take during their development seems to depend on where they mature. Lymphocytes that are to become B cells remain in the bone marrow, where they continue to develop. Those destined to differentiate into T cells migrate early in their development to the thymus, a small gland in the upper chest, behind the breastbone. (Figure 1 shows the different white cell lineages that differentiate from stem cells.) Together these lymphocytes provide the system with a potent two-pronged defense. Let us take a closer look at how they do their work.

B Cells

Mature B cells pump out protein molecules called *antibodies*. They do this in response to particular microorganisms, but only *after* they have come in contact with these microorganisms. Antibody production is what *vaccination* is all about. When you get a polio vaccine, you are receiving a weakened polio virus. The weakened germ will not cause infection but will stimulate the B cells. Once stimulated, B cells become minifactories, knocking out antibodies at the phenomenal rate of more than ten million an hour. That's almost three thousand antibody molecules a second.

Antibody molecules are very specific. A vaccine against polio will not protect you from tetanus or smallpox. The specificity is a result of a unique fit between an antibody molecule and certain surface molecules on the germs called *antigens*. Just as one key will fit only one lock, antibody molecules made by a particular B cell will fit only one type of antigen. Antibodies produced from weakened polio virus will fit only polio virus molecules.

This specificity baffled scientists for many years. Clearly there were millions of different antigens that the body had to produce antibodies against. The logical assumption was that antigens acted as templates. When a B cell wrapped itself around this template, it was instructed, or programmed, to produce the properly shaped antibodies.

Although this theory sounded good, it was false. Antibodies are protein molecules, and as such they are coded for by specific genes in the B cells. There seemed to be no way that a particular shape on an antigen could affect the type of protein the gene produced. Further investigation led to another theory of antibody formation called *clonal selection*, which is the currently accepted view. This theory postulates the existence of more than one hundred million different B cells, each one capable of recognizing a particular antigen, for which it then produces particular antibodies. In this theory the gene for making the antibody is already in place in the B cell before its encounter with the antigen and merely awaits activation.

The only problem with the theory is that one hundred million antibody proteins require one hundred million genes. Unfortunately, human cells have only one hundred thousand genes, most of which are not even involved with antibody production. Fortunately, the genes for making antibodies consist of several different segments. Contained in the human genetic material are many variations of each segment. Some segments have as many as three hundred different variations, all present in an immature B cell. As the cell matures in the marrow, it randomly picks and chooses segments, joining them together to form a unique, functional antibody gene. What we wind up with is a diverse population of B cells, each cell capable of recognizing and producing antibodies against one type of antigen. Recognition occurs through binding of antigen with antibody molecules that dot the surface of the B cell. (Remember, B cells are prolific antibody-making factories.)

Usually the binding of antibody to antigen does not destroy invading bacteria. Antibodies, as a rule, do not kill (although they can inactivate bacterial toxins and viruses) but rather call into action a class of murderous enzymes. These enzymes, numbering more than twenty, are collectively termed *complement* (for they complement antibodies), and they work by destroying outer membranes of bacteria. Once the killing and/or inactivation is accomplished, phagocytes are brought in to gobble up the mess. (Figure 2 summarizes the immune action of B cells.)

Figure 2
B-Cell Action

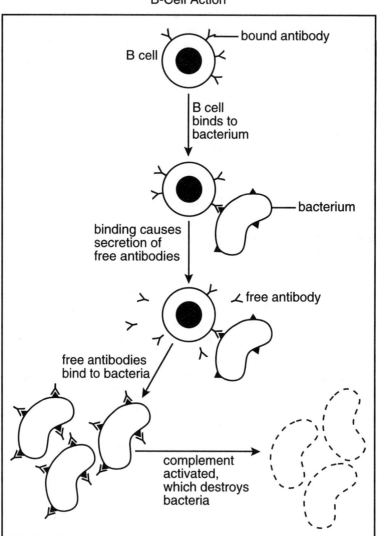

About ten trillion B cells are coursing through the blood and lymphatic system of a healthy person at any given moment. The antibodies they produce make up about one-fifth of the proteins found in blood. B cells and their antibodies constitute an arm of the immune system that is termed *humoral-mediated immunity*, *humoral* referring to the fluid nature of antibody secretions. By contrast, T cells bring about cellular-mediated immunity, for the key players in the drama are not humors or fluids but the cells themselves.

T Cells

A T cell earns its letter by migrating from the bone marrow to the thymus, where it subsequently matures. During the maturation process T cells, like B cells, become *immunocompetent*. This means they develop specificity—the ability to respond to only one type of antigen. But with T cells the story is considerably more complex. For starters, the thymus produces three very distinct populations of T cells. They are the *killer Ts*, the *helper Ts*, and the *suppressor Ts*. If we are interested in a good fight, the suppressor Ts are a disappointment. Triggered late in the immune response, they release chemicals that serve only to inhibit immune activity after the threatening microbes have been eliminated. Not so the killer and helper T cells. From the start, they are in the thick of the fray.

Killer and helper T cells do not respond to free bacteria, viruses, or other pathogenic agents lurking in our bodies. This is because, unlike B cells, they do not have the proper receptors on their cell surfaces to recognize these intruders. What they do recognize are our own body cells that have become infected with microorganisms. How they respond depends on whether they are killer T cells or helper T cells.

Killer Ts are the only T cells that kill. Their primary targets are body cells that harbor viruses, nasty little critters which become active only after penetrating living cells. Once inside a cell, viral particles busily employ the infected cell's machinery to make more copies of themselves. But the cell has evolved methods of preventing this viral piracy. Certain cellular proteins called *MHC* (major histocompatibility complex) molecules steal away snippets of viral protein and carry them to the cell membrane. Here they display themselves along with their abducted viral antigen to any and all killer T cells that might happen by. When the proper killer T arrives, it will bind with the MHC–viral antigen complex on the cell's surface. Then it will punch holes in the infected

Figure 3
Killer T-Cell Action

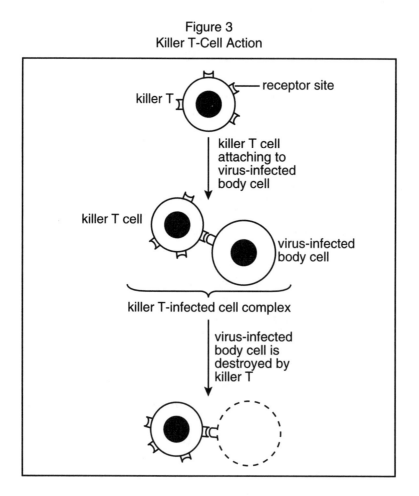

cell, enzymatically speaking. (Figure 3 summarizes the action of killer T cells.)

Like the bee that sacrifices its own life for the greater good by stinging an animal threatening its hive, cells infected with viruses serve themselves up to be slaughtered by killer T cells. In so doing, they prevent further viral multiplication and expose already existing viruses to B cells and their antibodies as well as phagocytes. Killer T cells also rid the body of cells that have turned cancerous and, unfortunately, are also involved in attacking and rejecting tissues from transplanted organs.

The system, up to this point, seems quite adequate and comprehensive. B cells produce antibodies, which, with help from phagocytes and complement, destroy free microbes. Killer T cells take care of any germs that might find their way into body cells. Although this battery makes helper T cells seem superfluous, this is not the case: it's the helper T cells that destroy the human immunodeficiency virus (HIV), which causes AIDS. Tell that to a dying AIDS patient however, and you'll learn a different viewpoint. It is the helper T cells that are destroyed by human immunodeficiency virus (HIV).

Helper T cells are the glue that holds the entire immune system together. Their secretions mobilize every other branch of the system, and they are anything but subordinate. Macrophages, B cells, and killer T cells are all turned on by interaction with helper T cells. It is done in much the same way killer T cells interact with virally infected body cells. Bits of viral protein are displayed on macrophage and B cell surfaces. Recognition and binding by helper Ts stimulate macrophages to digest their microbial meal and the B cells to proliferate and produce antibodies. (Figure 4 summarizes the immune action of helper T cells.)

In binding with various cells of the immune system, helper T cells also release a host of chemicals known as *cytokines*, or *interleukins*. These secretions further stimulate macrophage and B cell activity, in addition to sounding a

Figure 4
Helper T-Cell Action

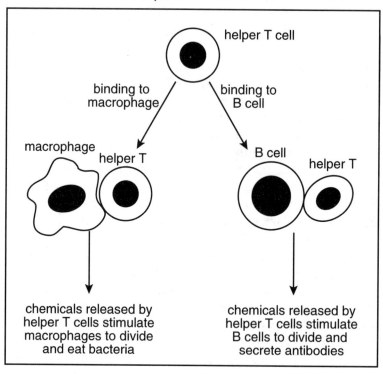

wake-up call to killer T cells. When all branches of the immune system are working efficiently, it is a surprisingly complex network of interacting cells and secretions, with the helper T cells right in the hub of it all.

Memory Cells

Twenty-five hundred years ago Thucydides of Athens, in writing of an epidemic disease, noted that the sick and dying were cared for by those who had recovered, "for no one was ever attacked a second time." This points out a fundamental property of the immune system, namely, that it confers immunity. Recover from mumps or the measles, and you can rest assured that you will not get it again. Vaccination is based on the same principle. Introducing a microbe into your body (even one weakened or dead) can confer lifelong protection against that disease. But why?

The answer lies in the creation of long-lived memory cells. When B cells respond to a first-time encounter with a particular virus or bacterium, they are converted into antibody-making machines. But this conversion takes more than five days, during which time the immune response is seriously compromised. Fortunately, though, some B cells (as well as T cells) are transformed into memory cells, which usually remain in the body for many years. Memory cells (B memory cells, that is) are primed to immediately flood the bloodstream with antibodies should a second encounter with the microorganism occur. So, assuming survival of a first assault, the body is well protected against reinfection. The only reason we keep getting diseases such as the common cold is that there are several hundred cold viruses, all sufficiently different from one another to prevent recognition by memory cells.

Over the billions of years life has been perfecting itself, it seems to have evolved a very adequate system of defense. The immune system, with its elaborate arsenal, is the second most complex system in the human body, surpassed only by the nervous system. There are, however, many chinks in the armor. Allergies, autoimmune diseases, and AIDS are all examples of an immune system that has, to a lesser or greater degree, failed. Why, in certain instances, is our immune system found so wanting? That is another story . . . and another essay.

Allergy: Fighting the Nonexistent Enemy

One day a young man sat down in a fine French restaurant in Minneapolis. He ordered steak served with a pale brown sauce called *French Nouve*. When it arrived, he ate the delicious dish with gusto. But, as Mark P. Freidlander, Jr., and Terry M. Phillips describe in their book *Winning the War Within*, his pleasure was short-lived: "After only a few minutes, he seemed to lose his breath. He wheezed twice, his mouth became numb and he began gulping for air. His hands gripped the edge of the table as he struggled to breathe." One dinner companion slapped him on the back. Another applied the Heimlich maneuver, believing he was choking on a piece of steak. Nothing helped, and, gasping for air, he lapsed into unconsciousness.

Luckily a paramedic who had rushed to the scene saw red blotches on the victim's forehead and cheeks and realized he was suffering an acute allergic reaction. As it turned out, the diner was allergic to shellfish. Unbeknownst to him, French Nouve sauce is made with, among other ingredients, puree of lobster. A shot of adrenaline soon opened up the man's air passages, and he quickly regained consciousness.

Such an allergic response, which is both systemic and potentially lethal, is termed *anaphylaxis*. Nuts, especially peanuts, are another food source prone to cause severe reactions in sensitive people, but the offending substance need not be a food ingredient. A stone tablet describes the death of Egyptian king Menes from an insect sting way back in 2641 B.C. This is, most likely, the first account of a fatality due to anaphylaxis. Unfortunately, there have been many others. It is estimated that several hundred to several thousand people in the United States die each year of anaphylaxis. Death usually results from shock or asphyxia.

For millions of others (20 to 25 percent of the population) the symptoms of allergy are, thankfully, much milder—runny nose, itchy eyes, sneezing, hives, and possibly gastrointestinal distress. These symptoms are nonetheless quite uncomfortable, and the cost of their relief runs into billions of dollars each year. Not surprisingly, the medical establishment has devoted much time and energy to uncovering the secrets of allergies.

An allergy occurs when our immune system is turned on in error. Normally the immune response is elicited by harmful microorganisms that invade the body. It protects us from infection. But for some reason, in people who are termed *hypersensitive*, harmless substances, called *allergens*, call the immune system to arms.

Outdoors, it is inhaling pollen that most often precipitates an allergic episode. Indoors, it is dust or, more specifically, the ugly eight-legged microscopic critters called *dust mites*. They thrive in dusty areas, feeding and traveling on microscopic flecks of sloughed-off skin (dander). The feces of these tiny relatives of the spider, when inhaled, evoke the sneezing-and-runny-nose syndrome referred to as *allergic rhinitis*.

Interestingly, cat allergy is not due to cat hair or cat dander, per se, but rather to a protein in the cat's saliva that it deposits on its fur while preening. Washing a cat once a month for three to eight months will cause it to stop making the saliva protein. This simple procedure, in effect, produces a nonallergenic cat.

Whatever the allergen, exposure to it precipitates a series of events culminating in an allergic episode—sometimes with alarming rapidity.

Sensitization

Only minutes after an allergen has been eaten, inhaled, or injected, symptoms appear. This is because the body of a susceptible person has already been primed by a previous encounter with the allergen—an encounter called *sensitization*. During the sensitization phase no allergic symptoms appear, but cellular activity goes on that will open the floodgates upon subsequent encounters. Sensitization can occur weeks, months, or even years before a second encounter triggers an allergic response.

Assuming the allergen has been inhaled, sensitization begins in the lining of the airways. Phagocytic white blood cells called *macrophages*, prowling the tissues, will devour and digest dust mite excreta or any other allergen. This represents the onset of an immune response. Bits of protein from the digested feces are then displayed on the surfaces of the macrophages, where they are recognized by certain *helper T white blood cells*.

Sound familiar? This is precisely what happens during a typical bacterial assault, discussed in the preceding essay, "Our Immune System." At this point in allergic sensitization, however, things proceed in a slightly different manner. Helper T cells release chemicals that stimulate *B white blood cells* to produce *antibodies*, just as they do in a microbial infection. But the antibodies produced during infection are of a different type from those released during an allergic response.

Antibodies can be divided into five different classes, based on their general structure and mode of action: immunoglobulin G (IgG), immunoglobulin M (IgM), immuno-

Figure 1
Sensitization

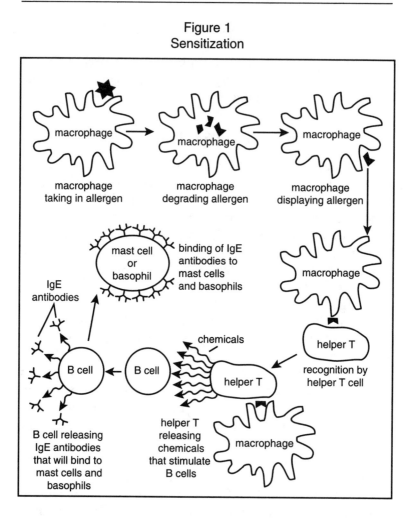

globulin A (IgA), immunoglobulin D (IgD), and immunoglobulin E (IgE).

IgG and IgM are the antibodies produced most commonly during normal viral and bacterial infections; rarely
do we see IgE antibodies. But this is not the case during
allergic sensitization, when IgE molecules are released in
huge quantities by the B cells. These Y-shaped antibody
molecules then bind, tail-first, to two types of cells: *mast
cells* and *basophils*. Mast cells derive from the bone marrow

and then migrate throughout the body, settling into skin, mucus membrane, and the tissue lining the lungs and gut. Basophils also derive from bone marrow but are a class of white blood cell. They are found circulating in the bloodstream. IgE antibodies have a particular affinity for both mast cells and basophils, and after sensitization they will dot the surfaces of these cells. (See Figure 1 for a detailed schematic of sensitization.)

The production of IgE antibodies, following first contact with an allergen, can take up to several weeks. During this time the allergen might be long gone, but the damage has already been done. Mast cells have been sensitized, the outstretched arms of their coating antibodies waiting patiently to ensnare the proper allergen—ready to spring into action as soon as a second encounter with the allergen occurs. That second encounter brings on stage two of an allergic response, *activation of mast cells.*

Activation of Mast Cells

Within seconds after reintroduction into the body, a particular allergen will bind to IgE molecules already in place on the mast cells. When an allergen molecule connects with

Figure 2
Activation of Mast Cells

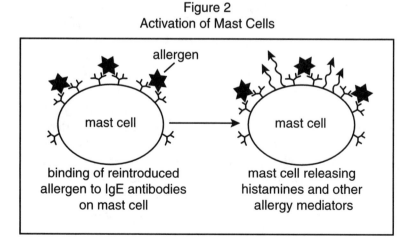

binding of reintroduced allergen to IgE antibodies on mast cell

mast cell releasing histamines and other allergy mediators

two neighboring IgE molecules, cross-linking them, there is a sudden, explosive release of chemicals called *allergic mediators* from the mast cells. It is these mediators that bring on the symptoms of allergy, and, if hypersensitivity is great enough, they can kill a person within minutes. (See Figure 2 for a detailed schematic of activation of mast cells.)

The most intensively studied and certainly the most infamous of all mediators is *histamine*. Ever since 1911 its nefarious role in the onset of allergic symptoms has been known. Antihistamines, which block histamine action, are a staple of allergy medications. There are, however, may other mediators, most notably the *prostaglandins* and *leukotrienes*. In the near future it is very possible that allergy medications that neutralize leukotriene action will prove just as valuable in combating allergy symptoms as the antihistamines.

As a group the allergy mediators do a number of things, none of them very pleasant. For starters, they dilate the small blood vessels and make them leaky. If the allergen remains contained and binds only with nearby mast cells, the sufferer will experience localized redness and swelling. This is what normally happens after a mosquito bite or bee sting. Nerve endings in the skin are also stimulated by the allergen (venom), producing the pain and/or itching.

Allergists take advantage of the inflammatory response to an allergen when they perform a skin test. In a typical test foreign substances are inoculated under the skin. If hypersensitivity to a particular inoculant exists, the area will redden and swell, termed a *wheal-and-flare reaction*.

Conditions, however, are not always as controlled as they are in a doctor's office. Sometimes the allergen does not remain contained but finds its way into the circulating blood. When this happens, things can get ugly. The allergen is carried to sensitized mast cells located throughout the body, initiating widespread mediator release. If fluid leakage from blood vessels into the tissue of the vocal cords becomes severe, the cords can swell to a point where they close off the trachea (windpipe). The man in the French restaurant suffered just such a fate. Fluid loss from blood vessels can also create a lethal drop in blood pressure. This is *anaphylactic shock*, a major cause of death in allergic overreaction.

Mediator activity also causes constriction of smooth muscle. This is the type of muscle found in the internal organs of the body. When constriction occurs in the muscle of the intestine, cramps and diarrhea can result. In the bronchial tubes of the lungs, constriction causes wheezing and breathing difficulty. Increased mucus secretion in the airways, another allergic symptom, only exacerbates the problem.

There is a curious link between allergy and asthma that is not completely understood. When someone suffers an asthma attack, bronchial tubes become constricted and breathing becomes labored. In the *extrinsic* form of the disorder an inhaled allergen triggers the attack, and allergic mediators are the culprit. But often asthma is *intrinsic*. In this form of the disease onset of an attack is not the result of an allergen. Strenuous exercise, emotional stress, and the breathing in of cold air can all trigger an attack of intrinsic asthma, and no one really knows why.

Late-Phase Reaction

The acute phase of an allergic response, mediated by mast cell activation, is immediate, can last up to an hour, and may be life threatening. Unfortunately, more times than not the allergic response does not end with this phase. After the acute phase subsides, there is frequently a return of symptoms in what is termed a *late-phase reaction*.

Laboratory work at Johns Hopkins University and elsewhere has demonstrated that it is primarily sensitized basophils (the white blood cells that become coated with IgE antibody molecules along with mast cells) that bring about and sustain the late-phase response. First they are attracted to sites of mast-cell activity, and then they are induced into secreting their own allergy mediators. Although the late-phase reaction does not take effect until several hours after the acute phase has died out, it is a more sustained reaction, and it can lead to chronic symptoms of allergy. Those persistent symptoms endured by some hay fever sufferers are undoubtedly due to ongoing and easily evoked late-phase inflammation.

In many respects an allergic response is the result of friendly fire—a situation in which soldiers are mistakenly fired upon by their own forces. In allergy, the body's own immune system, its army of cells and secretions, is erroneously called into action and does more harm than good.

One of the key players in this tragedy of errors is the inappropriately produced IgE antibodies. If doctors could somehow keep the B cells from spitting out these antibodies, they would avert an allergic response. This is the rationale behind allergy shots, which seek to desensitize allergy sufferers. In desensitization a patient is injected, over a period of months or years, with ever-increasing amounts of the problem-causing allergen. Like the old folk remedy for hangover, the patient is treated with a bit of the hair of the dog that bit him. The hope is that small doses of allergen will induce production of IgG antibodies. These IgG antibodies neutralize the allergen before it can trigger an IgE response.

Why Allergies?

At this point I would like to say a word or two on behalf of IgE antibodies. True, they are the perpetrators of allergy, but the human body is not so poorly evolved as to invest time and energy in the production of harmful or even merely useless proteins. Although not effective against bacteria and viruses, IgE antibodies are the ones of choice in combating larger parasites such as protozoa and microscopic worms.

The threadlike filarial worm is a good example of the type of parasite that invokes a strong IgE response. Spread by the bite of the mosquito, it is very prevalent in many tropical Third World countries. After the worm takes up residence in a human host, it multiplies rapidly. A drop of blood may contain thousands of the skinny, wriggling parasites. Eventually they block the lymphatics and cause extreme swelling in affected areas of the body. Legs often balloon to elephantine proportions—hence the name of the disease it produces, *elephantiasis.*

It is against invaders such as the filarial worm that many experts feel the IgE response has evolved. As soon as the parasite enters the body, huge numbers of IgE molecules are dispatched by the immune system—up to ten times those produced during an allergic response. Within fourteen days of the initial worm invasion each mast cell will become coated with several hundred thousand of these antibodies. At this point any contact with a worm will burst open the primed mast cells, flooding the body with their allergy mediators. These mediators bring about swelling and inflammation, which traps the parasite, preventing its spread.

But why are IgE antibodies called into action against strawberries and chocolate and penicillin and my mother-in-law's perfume? Is it merely because of a coincidental similarity in structure between their proteins and those of harmful parasites? Or is there a more purposeful design in the seeming aberrant functioning of the immune system that we, as yet, are unaware of? And why, if parasites induce the release of IgE antibodies, am I not allergic to my children?

Nobody truly understands the evolutionary significance of allergies, if indeed there is any. Its randomness bedevils researchers. Yet for all its unpredictability, allergies do tend to run in families. And David Marsh, a genetic allergist at Johns Hopkins, believes he has found a recessive gene that is at least partly responsible for allergic sensitivity. Its mode of action is as yet unclear. Will further research turn up other allergy-involved genes? Very possibly. Much has already been learned about allergic response on a cellular and molecular level. There is, however, so much more that awaits discovery.

Diamonds from Peanut Butter

In a 1957 episode of the TV show "Superman," the legendary Man of Steel takes a lump of coal in one of his bare hands and squeezes it with superhuman pressure while baking it with his x-ray vision. When he opens his hand a moment later, the coal is gone. In its place is a brilliant, sparkling diamond the size of a walnut.

There is, of course, no Superman, but the TV writers had the right idea about what it takes to make a diamond. They also understood how intrigued viewers would be by the possibility of creating one with such apparent ease. Diamonds have been treasured for thousands of years, in large part because of their nature but also because of their relative scarcity. Would the "Superman" viewers have been as impressed by the feat if they had known that it had been duplicated in a laboratory three years earlier?

To understand how it's possible for us to make, rather than just mine, diamonds, you need to know how they occur naturally.

Nature's Diamond Factory

The conditions necessary for diamond formation can be summed up in two words: *heat* and *pressure*. In the absence of Superman these conditions are not met naturally *on* Earth but *in* Earth. Diamonds are formed deep beneath Earth's surface—at least one hundred miles and perhaps up to three hundred miles down—in a layer of rock known as the *mantle*. At this depth deposits of carbon are subjected to extremes of heat and pressure for many millions of years, forming diamonds as large as pumpkins. Most diamonds are thought to be very old, forming up to *3.3 billion years ago*. They are, in fact, among the oldest minerals on Earth, sharing their beginnings with the earliest living things.

Diamonds are driven to Earth's surface through very long and narrow volcanic shafts, called *pipes*. These pipes are much longer and go down much deeper than those associated with conventional nondiamond-bearing volcanoes. It is believed that the journey of diamonds from Earth's interior to its surface takes less than a day—and that they erupt onto the surface with explosive force and without any flow of lava. No human has ever witnessed a diamond eruption.

The first diamonds were found more than four thousand years ago in stream beds, mixed with gravel and sand. Historically the richest deposits of diamonds have been in South Africa. The discovery of diamond fields there came about quite accidentally. In 1866 a farmer's child found "a pretty pebble" near the banks of a river. The pebble turned out to be a diamond worth $2,500. In 1979 a huge diamond deposit was found in western Australia. In fact the largest producer of natural diamonds in the world today is Australia. Zaire ranks second, with South Africa, Botswana, and Russia also major producers. (There are no commercial diamond mines in the United States, though diamond-bearing pipes have been discovered in the Canadian Northwest—the first of their kind in North America.) The world's diamond mines dig out nearly twenty-five tons of the precious gem each year.

Diamond prospecting, or locating areas profitable for mining, has not changed significantly over the years. Prospectors generally find diamond pipes by panning rivers for diamonds or indicator minerals, such as garnets. Geologists have some idea where to look, or where *not* to, but it is still largely a hit-and-miss process.

People the world over covet these ancient rocks like no other. What's the attraction?

The Nature of the Beast

We've been romancing the stone since the time of the ancients, who thought they were pieces of stars fallen from the heavens or bits of water frozen too long. In truth they are chunks of pure carbon, like the so-called lead in your pencil (which is not lead at all, but a form of carbon called *graphite*) and, yes, lumps of coal. What makes them different is the way in which the carbon atoms are put together.

Figure 1
Atomic Structure of Graphite

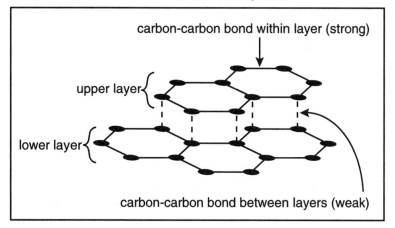

Atoms of carbon, you may remember from high school chemistry, bond four times with other atoms, including other carbon atoms. In graphite, the most common form of carbon, the atoms bond to form layers, or sheets. The carbon-carbon bonds within each layer are quite strong, but the bonds connecting one layer to another are much weaker (see Figure 1). These bonds are so weak, in fact, that the carbon layers are able to slide over one another easily. This causes graphite to have a slippery feel and makes it useful as a lubricant.

Diamond, on the other hand, has no weak bonds. Each atom is bonded to four other carbon atoms in a manner that forms the corners of a pyramid (see Figure 2).

Figure 2
Atomic Structure of Diamond

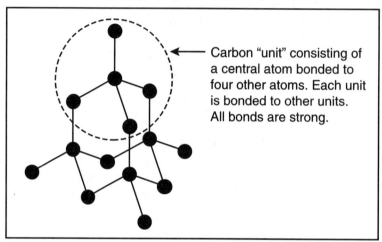

Carbon "unit" consisting of a central atom bonded to four other atoms. Each unit is bonded to other units. All bonds are strong.

The result is a tight network of strongly held atoms, giving diamond a very high melting point and an incredible hardness. (Diamonds contain more atoms per given volume than any other substance.) In fact diamond is the hardest substance that exists. It is five times harder than the next hardest natural mineral and nearly twice as hard as *anything* else, natural or otherwise. (With one possible exception: in July 1993 a substance was synthesized—carbon nitride (C_3N_4)—that is nearly as hard as diamond. When impurities are removed, scientists believe it may be harder.) It is also more brilliant and sparkling than anything else, because it slows down light more than any other transparent material. In a vacuum, light travels at its fastest: 186,000 miles per second. In water it slows down to about 140,000 miles per second. Window glass does an even better job, slowing it down to about 120,000 miles per second. In diamond light travels at *less than 77,000 miles per second*! That's 60 percent slower than it travels in a vacuum.

The importance of slowing light down is twofold. First, it causes the light to bend a great deal. (The more light slows in traveling from one transparent material into another, the more it bends.) As light passes into diamond, this bending, or *refraction*, causes it to bounce around a lot inside the diamond. A lot of bouncing creates a lot of places where the light can exit the diamond, giving it sparkle. Diamond is an incredible sparkler. Second, when light bends, it separates into a rainbow of colors—a process called *dispersion*. The more the light bends, the greater the dispersion. This results in the production of many different colors within the diamond. Diamonds not only sparkle; they dazzle with color.

Diamonds are also incredibly durable. They are impervious to the corrosive effects of salts and acids and other destructive chemicals. They do not erode easily and are therefore largely unaffected by the elements. In fact the word *diamond* comes from the Greek *adamas*, meaning "invincible." And they are nearly so. However, being pure carbon, diamonds will combine with oxygen at high temperature, disappearing entirely as carbon dioxide gas. Without oxygen, high temperature will cause diamond to revert to the more common form of carbon: graphite.

Dazzlingly radiant, imperviously hard and durable—it's no wonder that diamonds are highly prized.

Gemstones and Grinding Wheels

So highly valued are diamonds that they're the stuff of legends. Most people know the name of one of the world's famous diamonds, the Hope. The largest diamond ever found, however, was the *Cullinan*, named after the man who discovered the mine in South Africa from which it came. It weighed 3,106 carats, which is about 1⅓ pounds or .6 kilograms. (One carat equals two-tenths of a gram or seven-thousandths of an ounce. About 142 carats equal one ounce.) Since its discovery it has been cut into nine large and ninety-six smaller stones. The largest cut diamond in the world came from the *Cullinan*—the 530-carat *Star of Africa*. The second-largest cut diamond also came from the *Cullinan*.

Most diamonds of gem quality are transparent in color with a yellow tint. Colorless or blue-tinted diamonds are rarer and more valuable. Diamonds also come in red, green, orange, brown, and black, red being the rarest color. The highest price ever paid for a diamond, per carat, is $926,315.

Diamonds generally increase in value when they are *finished*. This involves cleaving or "cutting" the diamond and polishing it. Cutting gives the diamond shape and provides it with flat surfaces, called *facets*. The more common shapes include round, pear, emerald, and marquis (see Figure 3).

Most finished diamonds have fifty-eight facets. A stone that has been finished is generally more brilliant than an unfinished stone and provides a greater play of colors. The facets accomplish this by bouncing the light around more inside the stone. Diamond cutters are highly trained artisans; they may study a diamond for years before attempting to cut it.

The diamonds of legend are always gems, but in fact only a small percentage of the world's mined diamonds become jewelry. Most diamonds are small, not very transparent, and gray or brown in color. Their quality is not high enough for gemstones, but they are still very hard and durable, and industry has many uses for them. Most obvious perhaps is that they are used to cut other diamonds. They are also coated onto tools that cut and shape hard metal parts in making automobiles, trucks, trains, airplanes, and various types of machinery. They are pulverized into a fine

Figure 3
Common Shapes of Cut Diamonds
(size is approximately 1 carat each)

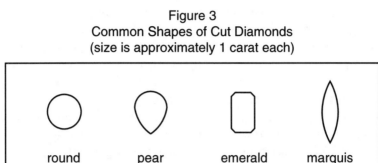

round   pear   emerald   marquis

grit or powder and used as an abrasive to grind and polish hard surfaces. They are coated onto drill bits and used to bore through rock for mining and for studying Earth's crust. They are used as needles in record players.

With the value of diamonds so high, their beauty and hardness unparalleled, and their uses manifold, it makes sense that we would try to duplicate nature—to make diamonds in the laboratory rather than wait for them to rise up and spew forth from the bowels of Earth. (Do not confuse *man-made* with *imitation* diamonds—such as cubic zirconium—which are not diamonds at all. Imitation diamonds do not approach the brilliance, durability, or hardness of true diamonds.)

The Pressure Is On

The thought is not a new one. Scientists have studied diamonds for centuries and have been trying to synthesize them since the early part of the nineteenth century. In 1797 an English chemist demonstrated that diamond is nearly pure carbon and differs from coal and graphite only in its external form. In 1828 several chemists claimed to have grown diamond crystals from solution—but they were mistaken. They had grown aluminum and magnesium oxide—the fool's gold of diamond makers. Others claimed to have made diamonds using an electric arc furnace or other heating devices. The problem with these early attempts is that they focused on temperature alone. Only after the discovery of diamonds in South Africa in 1866 and the extensive mining of diamonds that followed, did diamond makers realize the vital role of pressure in the diamond-making process.

A physics textbook defines pressure as force per unit area. In the British system it is measured in pounds per square inch. The pressure of the atmosphere at sea level is about 14.7 pounds per square inch. This pressure value is referred to as *one atmosphere*. Pressure is measured in many other units: torr, pascal, and bar, to name a few. For our purposes, however, the pressure unit atmosphere will do just fine.

The Problem with Diamonds

Why should it be so difficult to make a diamond? Why should it require such extreme conditions? It is pure carbon, give or take a trace of impurities—and carbon is as common as the air we breathe. Plastics are mostly carbon. Wood is mostly carbon. We are mostly carbon—except for water. An entire industry and branch of science, called *organic chemistry*, is based on carbon. There are more carbon compounds, by far, than all noncarbon compounds put together. So why are diamonds so rare—and so difficult to put together?

It has, once again, to do with the way they bond. The carbon bonds that form graphite are easy to make. They are low-energy bonds. The bonds that form diamond, on the other hand, require much more energy. The atoms must be squeezed together and subjected to high temperature to make the bonds stick. It's like rolling a ball up a hill. At the bottom of the hill the ball is graphite; at the top it is diamond. But it's not easy to get to the top. And it's not easy to stay there. Given sufficient temperature and pressure, however, the ball *will* roll up the hill and even find a place to sit so that it does not roll back down. Given sufficient temperature and pressure, we *can* make diamonds.

The Diamond Makers

Pressure was the big problem. At the turn of the century most laboratories involved in high-pressure research were unable to achieve sustained pressure greater than three thousand atmospheres. Beyond that point, the system either broke down or leaked. (Keep in mind that three thousand atmospheres is still considerable. The deepest parts of the ocean reach pressures just over one thousand atmospheres. But it was not nearly enough for diamond making.)

Percy Bridgman, an ingenious and inventive experimenter, took high-pressure research to the next level—to the next several levels. For half a century he devoted himself to this field. In 1905 he modified a piece of machinery used to generate high pressure and attained seven thousand atmospheres (that's *fifty tons per square inch*). In 1910 he got pressures greater than twenty thousand atmospheres. For the next two decades he subjected hundreds of compounds to these high pressures and discovered new properties in these materials and new crystalline forms. He got water to become ice at room temperature.

In the 1930s Bridgman set to work to improve the design of his high-pressure apparatus and ultimately achieved pressures in the range of four hundred thousand atmospheres—*nearly three thousand tons per square inch*! This pressure was certainly sufficient to make diamonds. For two years Bridgman placed sample after sample of graphite in his pressure cells in an attempt to form the precious gem. It didn't work. Either the pressure was not sustained, or the temperature was not there with it. Bridgman suspected that he was getting close but not close enough. It is now known that a pressure of at least fifty thousand atmospheres at a temperature greater than 1,800°F (1,000°C), for a sustained period of time is required to make diamonds.

For all his efforts, Percy Bridgman never made a single crystal of diamond. He was, however, awarded the Nobel

Prize in physics in 1946 in recognition of his lifelong achievement in the field of high-pressure research.

For the next chapter in the diamond-making saga we must travel across the Atlantic to Stockholm, Sweden, where diamond making was pursued in earnest by Sweden's major electric company, ASEA. The principal figure in the ASEA diamond project was Baltzar von Platen. Von Platen was called a "genius maniac"; he was an eccentric in the extreme and by some accounts quite mad. He planned to become famous inventing the first perpetual motion machine to solve the world's growing energy problem. To accomplish this he declared the second law of thermodynamics invalid. (This law states that energy is always lost as it does work and that machines cannot run forever without having that energy constantly replaced.)

In the laboratory, however, von Platen was remarkable. At ASEA he designed a diamond-making machine that produced pressures exceeding sixty thousand atmospheres. He was the inspiration for all who followed him at ASEA. One of those was Erik Lundblad, a large and imposing man known as a bon vivant. On February 16, 1953, he very quietly made history. He subjected graphite to a pressure of eighty-three thousand atmospheres at high temperature for a full hour and made the first synthetic diamond. The race to make diamonds was over.

Or was it? For inexplicable reasons the ASEA Corporation elected not to announce or publish or file a patent for its historic achievement. It remained silent. In the scientific community, for experimental results to be recognized they must be published and then duplicated by other scientists. Neither of these things happened. Robert Hazen, in his definitive book on diamond making, *The New Alchemists*, states, "Years after the event, von Platen regretted ASEA's failure to announce the creation of diamonds. By failing to share his secrets with the world, Baltzar von Platen lost his greatest opportunity for fame."

The race was still on!

For the final chapter of the diamond-making story we must cross the Atlantic again and return to America. In 1950 General Electric committed itself to making diamonds. The company that was the brainchild of Thomas Edison, that had pioneered countless inventions and improvements—including the long-lasting tungsten-filament bulb, fluorescent lighting, refrigerators, air conditioners, modern x-ray tubes, and electric trains—was now going to make the first diamonds.

It happened twice.

On the evening of Wednesday, December 8, 1954, research scientist Herb Strong loaded his high-pressure device with a sample of black carbon powder, raised the pressure to fifty thousand atmospheres and the temperature to 2,280°F (1,250°C), let it "cook" for sixteen hours (much longer than on his previous 150 attempts), and made two small diamonds.

On Thursday, December 16, 1954, another research scientist at General Electric, Tracy Hall, subjected a sample of graphite carbon to one hundred thousand atmospheres and 2,900°F (1,600°C) for thirty-eight minutes. Upon opening the high-pressure cell, Hall recalls:

> . . . my hands began to tremble. My heart beat wildly. My knees weakened and no longer gave support. Indescribable emotion overcame me and I had to find a place to sit down!
>
> My eyes had caught the flashing light from dozens of triangular faces of octahedral crystals . . . and I knew that diamonds had finally been made by man.

The race was over. But who won it—Strong or Hall? Logically, Strong should have won it; his experiment came first. However, his results were not reproducible, and reproducibility is critical to the confirmation and acceptance of any scientific claim. Hall's were reproducible. Also, experts have studied Strong's two crystals extensively in subsequent years and have decided that they could not possibly have been man-made. They are far too large, the shape and color are all wrong, and their x-ray patterns and infrared spectra show features found only in natural diamonds.

The first diamond, therefore, was created on December 16, 1954, by Tracy Hall in his laboratory at General Electric.

Since then, diamond making has become routine business. Diamonds can be made from just about *any* carbon-rich source. To quote again from *The New Alchemists*:

On a cold winter day in December 1955, Robert Wentorf, Jr., walked down to the local food co-op . . . and bought a jar of his favorite crunchy peanut butter. Back at his nearby General Electric lab he scooped out a spoonful, subjected it to crushing pressures and searing heat, and accomplished the ultimate culinary tour de force: he transformed the peanut butter into tiny crystals of diamond.

Diamonds have also been made from plastics, sugar, wood, roofing tar, and moth flakes, to name but a few. (In fact, this essay could have been titled "Diamonds from Moth Balls.")

Synthetic diamonds account for more than 80 percent of diamonds being used today. Their production exceeds one hundred tons per year. But they are primarily industrial, not the diamonds that adorn a finger, wrist, or earlobe. Although large diamonds of exceptional quality (greater than fourteen carats) have been made in the laboratory, the process is too expensive. It is best to leave gemstones to nature.

What's the Matter with Matter?

Yesterday, three scientists won the Nobel Prize for finding the smallest object in the universe. It turned out that it's the steak at Denny's.

Jay Leno

The year was 1990, and the three scientists Richard Taylor, Jerome Friedman, and Henry Kendall had been awarded Nobels for establishing, experimentally, the reality of quarks (the actual work had been done in the late 1960s). Quarks, as far as we know, are the most basic, fundamental, indivisible particles that make up matter. Their discovery is a fascinating bit of detective work that dates all the way back to 600 B.C. and to the musings of a number of ancient Greek philosophers.

The Ancient Greeks

The world is a very complex place, composed of millions— make that billions, no, trillions—of different substances. This type of complexity does not sit well with philosophers and scientists, who believe in an underlying simplicity. "Nature," said Albert Einstein, "though difficult to understand, ought to be simple and beautiful." Surely there must be a most basic substance or substances—building blocks of matter—that in combination produce all that there is.

Thales, circa 600 B.C., believed the basic substance to be water. And why not? It certainly was ubiquitous, and heating most substances released water. It was also one of the few substances that could be converted easily into any of the three phases of matter: solid, liquid, and gas. Still, Thales was wrong.

Then came Empedocles. He agreed with Thales's notion of simplicity but believed there were four fundamental elements: *earth, air, fire,* and *water,* the famous foursome later embraced by Aristotle and his followers. The almost infinite diversity of forms that matter could assume, he said, was the result of a mingling of these elements. For example, bone was composed of two parts earth, two parts water, and four parts fire. Empedocles also believed that forces were needed to meld the elements and came up with two beauties—*love* and *strife.* Aristotle also came up with two universal forces: *levity* and *gravity.*

By 420 B.C. or so, Democritus had come along. His claim to fame was the *atom,* or *atomos,* a tiny, discrete, indivisible particle that was as small as matter can get. Democritus conjured up the atomos (Greek for "not able to be cut") while smelling a freshly baked loaf of bread. How was the aroma reaching him? Simple. Atoms, too small to be seen, were zipping over to his nose from the bread. The atoms, however, were not exactly bread atoms. Rather, there existed a small number of differently shaped atoms that could be joined in various and sundry ways to form particles of bread or wine . . . or anything. Sweet things were made primarily of smooth atoms. Bitter things were composed of sharp atoms, liquids were composed of round atoms, and metals were formed by atoms that locked together.

Although scientists today do not subscribe to the variously-shaped-atoms concept, they marvel at Democritus's insights. With no instruments capable of investigating the makeup of matter, this ancient mathematician-astronomer came up with very sophisticated ideas. "Nothing exists," said Democritus, "except atoms and empty space; everything else is just opinion." He was right on the money.

Embracing the Atom

Unfortunately, the atom theory was not easily embraced, for it challenged the teachings of Aristotle, who believed matter to be continuous rather than particulate. The concept of continuous matter assumes that an object—a nail, for example—can be cut up into ever-smaller pieces, ad infinitum. The particulate theory, on the other hand, presupposes a smallest particle of iron, which can be cut no smaller.

Over the ensuing centuries not much happened to displace the old air, earth, water, and fire hypothesis. Careful study by astute men of science had expanded the list of elements to include salt, sulfur, mercury, oil, spirit, acid, alkali, and phlegm. (One could not forget phlegm.) The Aristotelian view of a nonparticulate universe held sway, but Democritus's notion of an indivisible atom would not go away. Chemists, for example, found that the behavior of a gas could best be explained if the gas was considered a vast collection of gas atoms. And the constant, never-changing weight ratios by which elements combined to form compounds (for example, hydrogen and oxygen combining to form water) suggested an atomic structure of the elements.

Many great scientists endorsed the atomist view of the world, as evidence for the corpuscular nature of matter continued to mount. Galileo and Newton were atomists. It seemed to make sense.

Finally, in 1808, John Dalton incorporated much of the prevailing scientific thought into a coherent theory. By that time chemists had discovered twenty-odd elements, which they assembled into a *periodic table* according to their weights. Dalton proposed that each element was composed of its own particular atom with its own particular weight.

These atoms joined together to form the molecules of compounds—each compound with its own unique mix of atoms. Water, for example, with a formula of H_2O, was composed of water molecules, each molecule consisting of two hydrogen atoms and one oxygen atom bonded together. At long last the elusive basic building blocks of matter had been nailed down. Well, not quite, but it was a start.

Dividing the Indivisible

Throughout the nineteenth century new elements continued to be discovered. By the late 1800s about fifty were known, each with its own distinctive, indivisible atom. Then experiments with cathode ray tubes began to shed doubt on the indivisibility of atoms. A beautifully simple model of matter's structure was about to go up in smoke, a manifestation of "the great tragedy of science," to quote Thomas Huxley, "the slaying of a beautiful hypothesis by an ugly fact."

A neon sign is a cathode ray tube. A television picture tube is a cathode ray tube. But in the nineteenth century a cathode ray tube was nothing more than a sealed length of glass that had been evacuated and filled with a particular gas. Each end of the tube had a metal fitting that allowed a battery to send an electric current through it. When the battery was connected properly, a bright, narrow beam traveled through the tube. This fascinated scientists, who hadn't a clue as to what comprised the bright beam.

Joseph John Thomson, working in the famous Cavendish Laboratory at Cambridge University, sought to find out. Using magnets and electrically charged plates to deflect the beam, he showed that it was particulate in nature, the particles having a specific mass and a negative charge. Surprisingly, no matter what gas was placed in the cathode ray tube, the mass and charge of the beam particle remained unchanged. It was obviously not composed of the gas atoms that filled the tube. And the mass of the particle was amazingly small, at least a thousand times lighter than the lightest of all atoms—hydrogen. J. J. Thomson had discovered the *electron*.

Scientists were compelled to accept the ugly fact that the atom was not indivisible. A smaller, more basic particle existed. If these negatively charged electrons were the stuff of atoms, they could not be the only stuff, for atoms were normally neutral and of much greater mass than the electron. A more massive and positively charged component of atoms must also exist. J. J. Thomson believed that the bulk of the atom was made of this large positively charged entity—a spongy or doughy blob—and that electrons were dispersed throughout it like plums in a plum pudding.

The model of the atom as an infinitesimal plum pudding, as sensible as it seemed, would be slain rather mercilessly by Ernest Rutherford in one of the all-time great experiments in science. Rutherford was a brilliant New Zealander who, at the turn of the century, did much work with a newly discovered phenomenon: radioactivity. It earned him a Nobel Prize in chemistry in 1908. It also set the stage for his greatest discovery, the *nucleus* of the atom. This is how it went:

Rutherford knew that radioactive substances emit particles of different types. One type is the *alpha particle*. It is emitted by substances such as radium. Alpha particles are fairly dense, positively charged missiles. If shot at a thin sheet of gold foil, they should go right through the puddinglike gold atoms. But gold atoms have positive and negative charges of their own, which should deflect the alpha particles somewhat. By noting these minor deflections, something of the nature of gold atoms might be ascertained.

To test his hypothesis, Rutherford set a small piece of radium in a lead container with a narrow hole. He aimed the hole at a thin sheet of gold foil. Completely encircling the foil was a zinc sulfide screen, which flashed brightly when struck by alpha particles. Now all he had to do was sit and watch where the flashes occurred on the screen.

Most of the alpha particles either went straight through

the foil or were deflected only slightly, as anticipated. But a tiny percentage of the particles (one in eight thousand) actually bounced off the gold foil. These occurrences astounded Rutherford, who would later claim, "It was quite the most incredible event that ever happened to me in my life. It was as if you fired a fifteen-inch artillery shell at a piece of tissue paper and it came back and hit you."

Once again the model of the atom had to be modified. By 1911 Ernest Rutherford had a new and improved version. The atom was not a spongy, positively charged blob studded with tiny electrons. Far from being embedded in a pudding, the electrons whizzed around a very small, very dense nucleus at the atom's center. J. J. Thomson's plum pudding had given way to Rutherford's miniature solar system, with only empty space between the nucleus and revolving electrons. Just how much space? Leon Lederman in his book *The God Particle* draws the following analogy: "To get a sense of the Rutherford atom, if we picture the nucleus as the size of a pea (about a quarter inch in diameter) the atom is a sphere of radius 300 feet."

Yes, that's right, a solid block of steel is 99.9 percent empty space. Only the interplay of electrical forces between subatomic particles creates the illusion of solidity.

Rutherford and his crew were not done yet. Continuing to bombard atoms with alpha particles, Ernest Marsden, one of Rutherford's students, discovered that the nucleus was comprised of yet smaller particles, which were almost two thousand times more massive than the electron and equal but opposite to it in charge. They were given the name *proton*.

The nucleus of a hydrogen atom contains only one proton. Other, heavier nuclei are created by clumping protons together.

In 1932, about a dozen years after the discovery of the proton, an English physicist, Sir James Chadwick, discovered a second type of particle inside the nucleus. Although almost identical to the proton in mass, it was electrically neutral, which accounts for its name: *neutron.* Its discovery seemed to complete the picture of the atom: a small, dense nucleus of heavy protons and neutrons with lightweight electrons revolving around it. Quantum theory would come into play to determine just how the electrons arrange themselves around a nucleus, but the important work had been done. The most fundamental, indestructible particles of matter had finally been uncovered. Right? Wrong. The discovery of yet more fundamental particles was still to come, and their story brings us to the modern era of physics.

Even More Fundamental Particles?

A key player in the discovery of the nucleus and its constituent particles was the radioactive element radium. To quote Robert Crease and Charles C. Mann in their book *The Second Creation*:

> The gram of radium in the Cavendish was at the heart of the quest. . . . It was kept inside a sort of oven made from lead bricks that rested in a skinny tower on the top floor of the laboratory. Low and heavy, the radium box was treated with the respect of an altar; only the trusted few were allowed near it.

Alpha particles, released during the decay of radium, bombarded the atomic nucleus and revealed its subatomic nature. But these particles, used so successfully by Rutherford and his students, were actually quite low in energy content. Although sufficiently energetic to crack open a fat nucleus and expose its contents, particles of far greater energy were needed to explore the structure of the protons and neutrons themselves—particles that would get their energies from *particle accelerators.*

These devices have only one purpose—to speed up charged particles, such as electrons and protons, which are then smashed into one another. The faster the particles move, the more energy they have, and the better they are at probing their target. The procedure is not particularly subtle. One physicist compared it to "smashing a pair of elegant Swiss watches together and then trying to figure out how they were designed by looking at the cogwheels and screws that would come flying away."

In the 1950s Robert Hofstadter, of Stanford University, aimed a beam of comparatively low-energy electrons at a vat of liquid hydrogen. The electrons bombarded the single-proton nuclei of the hydrogen atoms and were deflected or scattered. The pattern of scattering enabled scientists to determine not only the size of the hydrogen proton but also the distribution of its positive charge—a Nobel Prize-winning accomplishment.

In 1968 the experiment was repeated with a much more energetic beam of electrons, and a very different picture of the proton emerged. It was a more highly resolved picture in which, according to Leon Lederman, "three little guys were found to be running around inside the proton." Evidently even the proton was not solid, not indivisible, not the most fundamental particle.

These three little guys were the earliest experimental evidence of the reality of quarks. Several years earlier, Nobel Prize-winning theorist Murray Gell-Mann had conjured up quarks but believed them to be nothing more than mathematical constructs. That they "ran around" in threes reminded him of a line from *Finnegan's Wake*, by James Joyce: "Three quarks for Muster Mark." So he called the particle a quark. Most physicists at the time felt Gell-Mann's quarks made about as much sense as Joyce's. More about quarks in a moment.

Since the 1960s, many experiments have been run, colliding electrons with protons, protons with protons, protons with other subatomic particles such as antiprotons, and so on. The speeds of the particles and subsequent energies involved have climbed steadily. These high-energy collisions allowed scattering studies that have brought the pictures of elementary particles into sharper and sharper focus. But particle accelerators have done much more than this. Such collisions at near-light speed have accomplished another amazing feat—they have created new, never-before-seen particles . . . hundreds of them that scientists have dubbed the *particle zoo.*

How can this be? Isn't God the only one who can make a tree—or an elementary particle? God and particle accelerators. When highly energetic particles collide, their energies can be converted into mass, in accordance with Einstein's equation $e = mc^2$ (which says that *a lot* of energy is needed to create *a very little* mass; see "It's All Relative: The Special Theory"). More energetic collisions are capable of producing more massive particles. Most of these particles are unstable and survive for only a tiny fraction of a second. Quantum theorists believe that the void of space is awash with a multitude of these particles, which pop in and out of existence as sufficient energies collect momentarily. But why so many different particles? What ever became of the simplicity so beloved of physicists?

Enter the quark. Make that three quarks for Muster Gell-Mann. He showed that combinations of three different quarks—called *up, down,* and *strange*—could produce all or nearly all of the particles in the particle zoo. If he was correct, quarks were indeed the "atoms" that every physicist since Democritus had been searching for.

Gell-Mann *was* correct, but his model was incomplete. In the 1970s two more quarks were discovered, called *charm* and *bottom.* (Don't you just love these names?) Finally, in 1994, the sixth and hopefully last quark—the *top* quark—was detected at Fermilab, the world's most powerful particle accelerator. (Fermilab's particle accelerator, located an hour's drive from Chicago, Illinois, is a stainless-steel tube, narrower than a fire hose, that is bent into a circle four miles around and buried thirty feet beneath converted cornfields. It weaves through a thousand superconducting electromagnets that guide speeding protons and antiprotons into head-on collisions. Hooked up to the accelerator is a massive five-thousand-ton, three-story-tall collision detector facility. It is the largest machine any civilization has ever built.)

Today the six quarks are part of what is called the *Standard Model.*

The Standard Model

This is the triumph, the crowning achievement of twenty-five hundred years of physics. It is a theoretical edifice, confirmed by experiment, that, simply put, explains all that exists and all that ever has existed since the Big Bang. In short, it is a description of how the universe is put together. According to the Standard Model, matter is composed of two basically different classes of particles, called *quarks* and *leptons.*

Quarks

As already mentioned, there are six kinds of quarks—up, down, strange, charm, bottom, and top—whimsically termed *flavors.* (What ever happened to vanilla and chocolate?) The various flavors actually refer to quantum mechanical properties, since all quarks taste pretty much like chicken. (Each flavored quark comes in three different "colors," reflecting minor quantum differences.)

Quarks comprise the heavy or massive class of elementary particles. The top quark, in fact, is extraordinarily massive, weighing two hundred times that of a proton and nearly as much as an entire gold atom. That is why it was so hard to find, its creation (according to $E = mc^2$) requiring huge amounts of energy. Interestingly, only the up and down quarks are found in ordinary matter. The proton is composed of two ups and a down quark. The neutron has them in reverse, consisting of two downs and an up. The other flavors combine in groups of two or three to form the many exotic creatures of the particle accelerator's particle zoo. (High-energy radiation from space, called *cosmic rays*, creates unconventional particles naturally when it bombards Earth. Consisting mainly of protons, cosmic rays collide with atoms of the atmosphere, producing unorthodox, short-lived particles. The first antimatter particle—the positron, or antielectron—was discovered by studying cosmic ray bombardment.)

Quarks are never found alone, although solitary quarks probably did exist for less than a trillionth of a second, in the tremendous burst of heat and energy of the Big Bang.

Leptons

This is the class of light-mass elementary particles. As with quarks, there are six kinds of leptons. They include the *electron*, *muon*, and *tau* particle (all basically similar except for mass), as well as three totally massless particles, called *neutrinos*. Of all the leptons, only electrons are components of ordinary matter.

That's about it, unless, of course, you would like to know why quarks are always bound to other quarks. Or why a light bulb glows when it is turned on. Or why a baseball travels four hundred feet when struck by Don Mattingly's bat (I'm a Yankees fan). To understand these phenomena, we must take a closer look at yet another class of particles—the force-carrying particles. Yes, Alice, things keep getting curiouser and curiouser.

May the Force Be with You

There are four basic forces in the universe. We are familiar with two of them—*electromagnetism* and *gravity*—because they make things happen in our everyday world. At the subatomic level, however, two other forces are at work. They are the *strong force* and the *weak force*. The strong force binds quarks together and even holds protons and neutrons together in the nuclei of atoms. The weak force is responsible for radioactive decay and for the breakup of certain particles.

As far as the Standard Model is concerned, these forces are of interest because they are *all* mediated by elementary particles called *bosons*—different bosons mediating each force. The particles responsible for the strong force—what glues quarks together—are appropriately called *gluons*. Quarks seem to play a game of subatomic hot potato with the gluons, and this effectively binds them together. *Photons* are mediators of the electromagnetic force. When a heated object glows, it does so because energized electrons are throwing off photons. A magnet attracts a piece of iron because photons are exchanged between the two. And when a home run is smashed over the centerfield fence, it is an exchange of photons between atoms in the bat and the ball that repels the ball and sends it on its way.

The weak force is mediated by three types of particles: a positively charged W (W^+), a negatively charged W (W^-), and a Z. Although they are force and not matter particles, the Ws and the Z have mass, and quite a bit of it. Each weighs about eighty times that of a proton. W and Z particles, theorized in the 1950s, were not detected experimentally until 1983. It was only then that sufficient collision energies could be generated in particle accelerators to create such massive particles (remember, $e = mc^2$).

That brings us to gravity, the universal force of attraction that exists between everything exhibiting mass. It is a force 10^{41} times weaker than the electromagnetic force and 10^{41} times as big a headache. The mediator of gravity is a particle called the *graviton*, which has not yet been detected, perhaps because it is too massive, perhaps because it does not exist. Physicists hate gravity because they cannot bend it to their will. It is the only force whose equations refuse to fit neatly into quantum mathematics as the other three forces have. This bothers physicists, who believe that all the forces are related and were, in fact, one and the same force for the first trillionth of a trillionth of a trillionth of a second after the Big Bang. Cooling down separated out the forces as it did the matter particles. Whatever.

Gravity notwithstanding, the Standard Model is still alive and well. It is merely incomplete at the moment. And the graviton is not the only missing particle. Many physicists believe there is a mysterious fellow called the *Higgs boson* that interacts with various other particles, giving them the property of mass. Nobel laureate Leon Lederman calls Higgs the "God particle" and feels its discovery is essential to an understanding of God, the universe, and everything.

The *superconducting supercollider* (SSC) was supposed to find Mr. Higgs. It would have been a fifty-four-mile-around particle accelerator several times more powerful than any presently existing. But its $11 billion price tag was too dear for the U.S. Congress, and in October 1993 the project was scrapped. Perhaps you remember the event. It was the day particle physicists began accelerating as they jumped out of windows in dismay.

Einstein sought what he called "a simple and lucid image of the world—a single principle that would account for the baffling differences between the forces and the great variety of particles." Our current understanding of the cosmos has, in a sense, failed in this regard. Even at its most basic level there are six quarks, six leptons, and a half a dozen force particles. Each quark comes in three different colors, there are eight varieties of gluons, and every matter particle has its antiparticle. That makes a total of sixty fundamental particles. Will the future reveal yet tinier, more fundamental "atoms" of which these sixty are composed? Physicists do not believe so. I guess they've never eaten at Denny's.

Index